Oral Microbiology

Third edition

Philip Marsh
Public Health Laboratory Service Centre for
Applied Microbiology and Research
Salisbury

and

Michael Martin
Department of Clinical Dental Sciences
University of Liverpool

CHAPMAN & HALL
University and Professional Division
London · Glasgow · New York · Tokyo · Melbourne · Madras

Published by Chapman & Hall, 2-6 Boundary Row, London SE1 8HN, UK

Chapman & Hall, 2-6 Boundary Row, London SE1 8HN, UK

Blackie Academic & Professional, Wester Cleddens Road, Bishopbriggs, Glasgow G64 2NZ, UK

Chapman & Hall GmbH, Pappelallee 3, 69469 Weinheim, Germany

Chapman & Hall USA., One Penn Plaza, 41st Floor, New York, NY10119, USA

Chapman & Hall Japan, ITP - Japan, Kyowa Building, 3F, 2-2-1 Hirakawacho, Chiyoda-ku, Tokyo 102, Japan

Chapman & Hall Australia, Thomas Nelson Australia, 102 Dodds Street, South Melbourne, Victoria 3205, Australia

Chapman & Hall India, R. Seshadri, 32 Second Main Road, CIT East, Madras 600 035, India

First edition 1980
Second edition 1984
Reprinted 1985, 1988, 1989
Third edition 1992
Reprinted 1992, 1994

© 1980, 1984, 1992 P.D. Marsh and M.V. Martin

Typeset in 10/12pt Palatino by Intype, London
Printed in Great Britain by St Edmundsbury Press, Bury St Edmunds, Suffolk

ISBN 0 412 43360 5

A Catalogue record for this book is available from the British Library

Library of Congress Cataloging-in-Publication Data available

To Jane, Katherine, Thomas and Jonathan

Contents

x Contents

Preface to the third edition

Since the second edition of *Oral Microbiology* was published in 1984, there has been an enormous expansion in knowledge across the complete breadth of this subject area. This has resulted in the need to not just update but to totally rewrite the previous edition. In particular, there have been striking advances in the taxonomy of the oral microflora, which is enabling closer associations between certain organisms with health and disease to be discerned. These changes in classification have caused difficulties when comparing new studies with earlier work when a different nomenclature was in use. It is hoped that the comparisons of the past and present classification schemes presented in this edition will be of value to young and experienced researchers alike. The significance of the role of endogenous substrates to the nutrition of the resident oral microflora, and the complex relationship between the host defences and the normal flora is also discussed extensively in this edition.

The last few years have seen renewed interest and progress in the understanding of dental caries, particularly of the root surface in the elderly. Advances have continued to be made in the microbial aetiology of periodontal diseases, while the precise role of specific bacterial species in tissue breakdown remains a contentious area. These advances in knowledge are being converted into improved methods of diagnosis, treatment, and prevention. New research into oral manifestations of AIDS, the control of cross infection and of other oral infections has made this an opportune time to write an enlarged, third edition of this book.

It is intended that this new edition will continue to help science and clinical students of all levels, as well as research workers and teachers alike, to understand the microbial ecology of the oral cavity and its importance in health and disease.

We would like to thank our colleagues who have helped and advised us, especially Else Theilade, Michael Pang, Jeremy Hardie, Rob Whiley, Saheer Gharbia, Haroun Shah, Roy Russell and Pauline Handley. We would also like to thank the following individuals and publishers for granting permission to reproduce data or figures: Alan Dolby (Figure 6.2) and Pauline Handley (Figure 4.5, Table 4.6); American Society for Microbiology (Figure 4.5); Cambridge University Press (Figure 7.3, Table 7.7); Harwood Academic Publishers (Table 4.6); Journal of Dental Research (Tables 6.9 and 6.10); and MTP Press Ltd (Figures 2.6 and 4.2, Table 6.1). Particular thanks also go to our families who have put up with so much during the preparation of this book.

P. D. Marsh, *Salisbury*
M. V. Martin, *Liverpool*

Preface to the second edition

Oral microbiology forms an important part of the curriculum of dental students while the multidisciplinary nature of the research in this area means that studies of the adherence, metabolism and pathogenicity of oral bacteria are equally relevant to microbiologists. The success of the first edition of *Oral Microbiology* stems in part from the fact that the book satisfies successfully the needs of both of these groups of students as well as those of general dental practitioners, medical students and senior scientists.

Since the publication of the first edition there have been rapid advances in our knowledge of the mechanisms involved in dental plaque formation, in the metabolism of oral bacteria both in health and disease, and in the aetiology of periodontal disease. This knowledge is being applied to the development of specific measures to control the incidence and severity of oral infections. Consequently, the second edition has been enlarged to include two comprehensive chapters on the current theories of the aetiology and prevention of these diseases. Elsewhere in the second edition, recent advances in our understanding of the colonization of the tooth surface by oral bacteria and in the biochemical basis of their pathogenicity have also been incorporated, while the glossary of microbiological and dental terms has been extended. The nomenclature of oral bacteria has been brought up to date, although this is an area in constant flux and the future will result in minor changes in the bacterial nomenclature as used in this text.

We would like to thank Derek Ellwood, Bill Keevil, Ailsa McKee, Haroun Shah and Geoffrey Craig and all our colleagues who have helped and advised us in the preparation of the second edition. Especial thanks go to our wives, Jane and Anne, and our families for their patience, support and encouragement.

P. D. Marsh, *Salisbury*
M. V. Martin, *Liverpool*

1984

Preface to the first edition

Micro-organisms in the mouth are responsible for a variety of infections including the most prevalent diseases affecting man today. A vast amount of pure and applied research in many disciplines is underway in an attempt to determine the aetiology of these oral infections with the hope that preventive measures can be developed. Results from such studies appear in a bewildering spectrum of journals and although review articles on specific topics have been published there have been only a few attempts to produce a text-book devoted solely to oral microbiology. The object of this book is to bridge this gap in the literature by providing a general introduction to the microbiology of the mouth both in health and disease.

Oral Microbiology has been written deliberately to appeal also to students of dentistry who lack an inexpensive text dealing specifically with this important part of their syllabus. Naturally, terminology will have been used that is unfamiliar to students of either subject. I hope that the provision of glossary will overcome any problems that might arise.

A book of this length produces severe constraints on the amount of material that can be included. Of necessity, a few topics have had to be omitted or treated superficially and some areas of current debate and controversy have been simplified for clarity.

I am indebted to many colleagues for their constructive criticism and advice during the preparation of the manuscript, particularly Dr George Bowden, Dr Jeremy Hardie and Dr Paul Rutter. All the photographs were kindly provided by Mr Alan Saxton of Unilever Research. I would like to take this opportunity to thank the authors and publishers who gave their permission for me to reproduce tables and figures from their articles. In addition I want to express my gratitude to Mrs Mabel Cockerill for her excellent and accurate typing of the manuscript, to Dr Dominic Recaldin and his staff of Thomas Nelson and Sons Ltd., and to the series' editors for their help in making the writing of this, my first book, such a painless experience. Finally, I would like to thank my wife, Jane, for her support and encouragement during all stages of the preparation of this book.

London, 1979   P. D. Marsh

1 Oral microbiology. Background and introduction

It has been estimated that the human body is made up of over 10^{14} cells of which only around 10% are mammalian. The remainder are the micro-organisms that comprise the resident microflora of the host. This resident microflora does not have merely a passive relationship with the host but contributes directly and indirectly to the normal development of the physiology, nutrition and defence systems of animals and humans. The microbial colonization of all environmentally-exposed surfaces of the body (both external and internal) begins at birth. Such surfaces are exposed to a wide range of micro-organisms derived from the environment and from other humans. However, each surface, because of its physical and biological properties, is suitable for colonization by only a proportion of these microbes. This results in the acquisition and natural development of diverse but characteristic microfloras at particular sites. In general, these microfloras live in harmony with the host and, indeed, both parties benefit from the association.

THE ROLE OF THE ORAL MICROFLORA IN HEALTH AND DISEASE

The mouth is similar to other sites in the body in having a natural microflora with a characteristic composition and existing, for the most part, in a harmonious relationship with the host. However, perhaps more commonly than elsewhere in the body, this relationship can be broken and disease can occur in the mouth. This is usually associated with:

(a) Major disturbances to the habitat which perturb the stability of the microflora. These disturbances can be from exogenous sources (e.g. following antibiotic treatment or the frequent intake of fermentable carbohydrates) or they can be derived from endogenous changes (e.g. alterations in the integrity of the host defences).
(b) The unexpected presence of bacteria at sites not normally accessible

to them (e.g. following tooth extraction or other trauma, when oral bacteria enter tissues or the blood stream and are disseminated around the body).

Bacteria with the potential to cause disease in this way are termed 'opportunistic pathogens', and many oral organisms have the capacity to behave in this manner. Indeed, most individuals suffer at some time in their life from localized episodes of disease in the mouth caused by imbalances in the composition of their resident oral microflora. The commonest clinical manifestations of such imbalances include dental caries and periodontal diseases, both of which are highly prevalent in industrialized societies, and are on the increase in developing countries. Dental caries is the dissolution of enamel (demineralization) by acid produced primarily from the metabolism of dietary carbohydrates by the bacteria attached to the tooth surface in dental plaque. Dental plaque is also associated with the aetiology of periodontal diseases. Periodontal diseases are a group of disorders in which the supporting tissues of the teeth are attacked; this can eventually lead to the loss of the tooth. Although rarely life-threatening, the direct cost of treating these common oral diseases is enormous and they are the most expensive infections that the majority of individuals will have to contend with in their lifetime. In 1984, in the USA, the cost of dental treatment was estimated at approximately 24 billion dollars while the cost to the National Health Service of the United Kindom in 1986–87 was over £750 million. In addition, dental treatment and disease is associated with pain and trauma, and results in a huge loss of time from work. With people retaining their teeth for longer periods due to scientific and clinical advances, the number of treatment courses (and hence costs) continue to rise. Some idea of the prevalence of dental disease in England and Wales (most of which is mediated by micro-organisms) can be gauged by the numbers of treatment items provided for these areas by the National Health Service (Table 1.1). Not surprisingly, therefore, the main thrust of research in oral microbiology is directed towards understanding the processes involved in the two major groups of dental diseases: caries and periodontal diseases.

Dental caries and periodontal diseases result from a complex interaction of diet, the resident microflora, and the host. In order to determine the mechanisms behind these diseases it is necessary to understand the ecology of the oral cavity. In this approach the relationships among micro-organisms, and between the micro-organisms and the host, will be determined. The basic composition of the oral microflora is reasonably well known but less is understood of the mouth as

Table 1.1 Prevalence of dental disease in England and Wales as indicated by the items of treatment provided by the National Health Service, 1986–87

Treatment item	Number
Restorations	25 000 000
Crowns	2 500 000
Extractions	4 500 000
Root canal therapy (endodontics)	1 500 000
Bridges and dentures	1 500 000
Treatment courses	34 000 000

an environment, and how the properties of this environment influence the composition and metabolism of the resident microflora.

TERMINOLOGY

Much of the terminology used in this book to describe events in microbial ecology will be as defined by Alexander (1971). The site where micro-organisms grow is the **habitat**. The micro-organisms growing in a particular habitat constitute a **microbial community** made up of populations of individual species or less well-defined groups (taxa). The microbial community in a specific habitat together with the abiotic surroundings with which these organisms are associated is known as the **ecosystem**. The most mis-used ecological term of all, the **niche**, describes the function of an organism in a particular habitat. Thus, the niche is not the physical position of an organism but is its role within the community. This role is dictated by the biological properties of each microbial population. Species with identical functions in a particular habitat will compete for the same niche, while the co-existence of many species in a habitat is due to each population having a different role (niche) and thus avoiding competition.

A number of terms have been used to describe the characteristic mixtures of micro-organisms associated with a site. These include the normal, indigenous, or commensal microflora. Some authors have argued that the term **commensal** should not be used because it implies a mutually neutral or passive relationship between micro-organism and host. This is patently not true since the mucosal surfaces of germ-free animals are anatomically and physiologically different from their conventional counterparts, while the host defences undoubtedly inter-act in some way with the microflora. Moreover, many oral organisms can act as opportunistic pathogens and so the inclusion of such species within the normal or commensal microflora can create problems of semantics. The term **amphibiont** has been suggested as a replacement term, signifying a spectral position between symbiosis and pathogenicity, and merging with both. The typical amphibiont would be, however, neither obligately pathogenic nor non-pathogenic under all circumstances. This would avoid the arguments over whether potentially pathogenic micro-organisms such as certain yeasts, which can be carried asymptomatically by some individuals, should be regarded as members of the normal microflora of a site. However, the term amphibiont has not been widely used. Alexander (1971) proposed that species found characteristically in a particular habitat should be termed **autochthonous** micro-organisms. These multiply and persist at a site and contribute to the metabolism of a microbial community (with no distinction made regarding disease potential), and can be contrasted with **allochthonous** organisms which originate from elsewhere and are generally unable to colonize successfully unless the ecosystem is severely perturbed. Alternatively, a simple approach has been to use the term **resident microflora** to include any organism that is regularly isolated from a site. Again, no distinction concerning disease potential is made.

Micro-organisms that have the potential to cause disease are termed **pathogens**. As stated earlier, those that cause disease only under exceptional circumstances are described as **opportunistic pathogens**, and can be distinguished from **frank pathogens** which are associated consistently with a particular disease.

The properties of the mouth that influence its function as a microbial habitat together with the major groups of micro-organisms that reside there will be described in the next two chapters. Subsequent chapters will describe the acquisition and development of the oral microflora (Chapter 4), especially dental plaque (Chapter 5), while the final chapters will consider the role of the oral microflora in disease.

SUMMARY

The mouth has a resident microflora with a characteristic composition that exists, for the most part, in harmony with the host. Components of this microflora can act as opportunistic pathogens when the habitat is disturbed or when micro-organisms are found at sites not normally accessible to them. Dental diseases, caused by imbalances in the resident microflora, are highly prevalent and extremely costly to treat.

FURTHER READING

Alexander, M. (1971) *Microbial Ecology*. John Wiley and Sons Inc., New York.

Lynch, J. M. and Poole, N. J. (1979) *Microbial Ecology. A conceptual approach*. Blackwell Scientific Publications, Oxford.

Rosebury, T. (1962) *Micro-organisms Indigenous to Man*. McGraw-Hill, New York.

2 The mouth as a microbial habitat

The ecological properties of the mouth make it different from all other surfaces of the body. However, the mouth must not be regarded as a uniform environment. It consists of several distinct habitats each of which will support the growth of a characteristic microbial community. The habitats of the oral cavity that provide obviously different ecological conditions for colonization and growth include the lips, cheek, palate, tongue, gums and teeth. The properties of these habitats will change during the life of an individual. For example, during the first few months, the mouth consists only of mucosal (epithelial) surfaces for microbial colonization. With the development of the primary dentition, hard non-shedding surfaces appear providing a different type of surface for the deposition and growth of oral micro-organisms. The eruption of teeth also generates another habitat, the gingival crevice, where the tooth rises out of the gums, (Figure 2.1 and 2.2) with its own particular nutrient source (gingival crevicular fluid, GCF).

Ecological conditions within the mouth are never stable for long periods. The oral ecosystem will be subjected to considerable variation during the change from primary to permanent dentition. In addition, the ecology of the mouth will be affected by extraction of teeth, the insertion of prostheses such as dentures, and any dental treatment including scaling, polishing and fillings. Transient fluctuations in the stability of the oral ecosystem may be induced by the frequency and type of food ingested, variations in saliva flow, and periods of antibiotic therapy.

The unique nature of the mouth as a habitat has been emphasized. Four features that help make the oral cavity distinct from other areas of the body: teeth; specialized mucosal surfaces; saliva and gingival crevicular fluid will be considered now in detail.

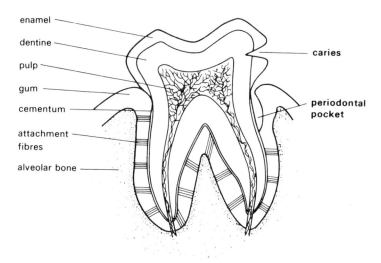

enamel

dentine

pulp

gum

cementum

attachment
fibres

alveolar bone

caries

periodontal
pocket

Figure 2.1 Tooth structure in health and disease.

Teeth

The mouth is the only site in the body that has hard non-shedding surfaces for microbial colonization. These unique tissues allow the accumulation of large masses of micro-organisms (predominantly bacteria) and their extracellular products. This accumulation is called dental plaque and, while it is found naturally in health, it is also associated with dental caries and periodontal diseases. The properties of dental plaque will be described in detail in Chapter 5, and its relationship to disease will be discussed in Chapters 6 and 7.

Each tooth is composed of four tissues – pulp, dentine, cementum and enamel (Figure 2.1). The pulp receives nerve cells and blood supplies from the tissues of the jaw via the roots. Thus the pulp is able to nourish the dentine and act as a sensory organ by detecting pain. Dentine makes up the bulk of the tooth and functions by supporting the enamel and protecting the pulp. Enamel is the most highly calcified tissue in the body and is normally the only part of the tooth exposed to the environment. Cementum is a specialized calcified connective tissue that covers and protects the roots of the tooth. Cementum is important in the anchorage of the tooth, being attached to fibres from the alveolar bone of the gum. With ageing, recession of the gingival tissues can occur exposing cementum to microbial colonization and attack. Root-surface caries is a consequence, and is an increasing problem in the burgeoning elderly population (Chapter 6).

Teeth do not appear in the mouth until after the first few months of

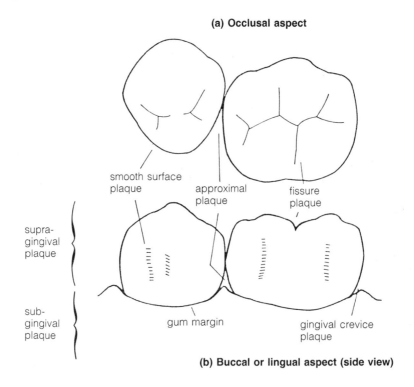

(a) Occlusal aspect

smooth surface plaque

approximal plaque

fissure plaque

supra-gingival plaque

sub-gingival plaque

gum margin

gingival crevice plaque

(b) Buccal or lingual aspect (side view)

Figure 2.2 Diagram illustrating the different surfaces of a tooth, and the terminology used to describe plaque sampling sites.

life. The primary dentition is usually complete by the age of three years, and around six years the permanent teeth begin to erupt. This process is complete by about 12 years of age. During these periods of change local ecological conditions will vary, which will in turn influence the resident microbial community at a site.

The ecological complexity of the mouth is increased still further by the range of habitats associated with the tooth surface. Teeth do not provide a uniform habitat but possess several distinct surfaces (Figure 2.2), each of which is suitable for colonization only by certain populations of bacteria. This is due to the physical nature of the particular surface and the resulting biological properties of the area. The areas between adjacent teeth (approximal) and in the gingival crevice afford protection from the adverse conditions in the mouth. Both sites are also anaerobic and, in addition, the gingival crevice region is bathed in the nutritionally-rich GCF so that these areas support a more diverse community. Smooth surfaces are more exposed to environmental forces and can be colonized only by a limited number of bacterial species adapted to such extreme conditions. The properties of a smooth surface will differ according to whether it faces the cheek (buccal surface) or the inside (lingual surface) of the mouth. Pits and fissures of the biting (occlusal) surfaces of the teeth also offer protection from the environment. Such protected areas are associated with the largest microbial communities and, in general, the most disease.

The relationship between the environment and the microbial community is not unidirectional. Although the properties of the environment dictate which micro-organisms can occupy a given site, the metabolism of the microbial community can modify the physical and chemical properties of their surroundings. Thus, the environmental conditions on the tooth will vary in health and disease (Figure 2.1). As caries progresses, the advancing front of the lesion penetrates the dentine. The nutritional sources will change and local conditions will become acidic and more anaerobic due to the accumulation of products of bacterial metabolism. Similarly, in disease, the gingival crevice develops into a periodontal pocket and the production of gingival crevicular fluid is increased. These new environments will select the microbial community most suitably adapted to the prevailing conditions.

Mucosal surfaces

Although the mouth is similar to other ecosystems in the digestive tract in having mucosal surfaces for microbial colonization, the oral cavity does have specialized surfaces which contribute to the diversity of the microflora at certain sites. The papillary structure of the dorsum of the tongue provides refuge for many micro-organisms which would otherwise be removed by mastication and the flow of saliva. The tongue can also have a low redox potential, which enables obligately anaerobic bacteria to grow. Indeed, the tongue may act as a reservoir for some of the Gram-negative anaerobes that are implicated in the aetiology of periodontal diseases (Chapter 7). The mouth also contains keratinized (as in the palate) as well as non-keratinized stratified squamous epithelium which may affect the intra-oral distribution of micro-organisms.

Saliva

The mouth is kept moist and lubricated by saliva which flows over all the internal surfaces of the oral cavity. Saliva enters the oral cavity via ducts from the major paired parotid, submandibular and sublingual glands as well as from the minor glands of the oral mucosa (labial, lingual, buccal and palatal glands) where it is produced. There are differences in the chemical composition of the secretions from each gland, but the complex mixture is termed whole saliva. Saliva contains several ions including sodium, potassium, calcium, chloride, bicarbonate and phosphate (Table 2.1); their concentrations vary in resting and stimulated saliva. Some of these ions contribute to the buffering property of saliva which can reduce the cariogenic effect of acids produced from the bacterial metabolism of dietary carbohydrates. Although the bulk secretion rate for whole saliva is in the range of $0.5-111.0$ ml h^{-1} (mean $= 19$ ml h^{-1}), saliva flows as a thin film (0.1 mm deep) over oral surfaces. This flow is slowest in the upper anterior buccal region (0.8 mm min^{-1}) and fastest in the lower anterior lingual region (8.0 mm min^{-1}), and this may influence the caries susceptibility of a site.

Table 2.1 A comparison of the mean concentration (mg/100 ml) of some constituents of whole saliva and gingival crevicular fluid (GCF) from humans

	Whole saliva		
Constituent	Resting	Stimulated	GCF
Protein	220	280	7×10^3
IgA	19		110*
IgG	1		350*
IgM	<1		25*
C$_3$	tr	tr	40
Amylase	38		–
Lysozyme	22	11	+
Albumin	tr	tr	+
Sodium	15	60	204
Potassium	80	80	70
Calcium	6	6	20
Magnesium	<1	<1	1
Phosphate	17	12	4
Bicarbonate	31	200	–

tr = trace amounts
* determined in GCF samples from patients with periodontitis.

Bicarbonate is the major buffering system in saliva but phosphates, peptides and proteins are also involved. The mean pH of saliva is between pH 6.75–7.25, although the pH and buffering capacity will vary with the flow rate. Within a mouth the flow rate and the concen-

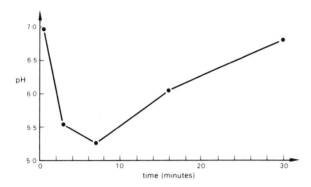

Figure 2.3 Stephan curve. Typical response of plaque pH to fermentable carbohydrate.

tration of components such as proteins and calcium and phosphate ions have circadian rhythms, with the slowest flow of saliva occurring during sleep. The major organic constituents of saliva are proteins and glyco-proteins, such as mucin. These glycoproteins influence (a) the aggregation and adhesion of bacteria to oral surfaces (Chapter 4); (b) they interact with other components of the host defences; and (c) they act as the primary sources of nutrients (carbohydrates and proteins) for the resident microflora. Other nitrogenous compounds provided by saliva include urea and numerous free amino acids. Not all of the amino acids essential for the growth of oral bacteria are present, and these may be derived from the enzymic breakdown of salivary proteins and peptides. The concentration of free carbohydrates is low in saliva, and most have to be derived from the metabolism of glycoproteins.

The metabolism of amino acids, peptides, proteins and urea can lead to the net production of base, which contributes to the rise in pH following acid production after the dietary intake of fermentable carbohydrates (Figure 2.3). In particular, a salivary tetrapeptide with the sequence gly-gly-lys-arg (termed sialin) can be converted to ammonia and putrescine by oral bacteria and can cause a pH-rise effect in the presence of low sugar concentrations.

Several anti-bacterial factors are present in saliva (Table 2.2) which are important in controlling bacterial and fungal colonization of the mouth, and include lysozyme, lactoferrin, and the sialoperoxidase system. Antibodies have been detected, with secretory IgA being the predominant class of immunoglobulin; IgG and IgM are also present but in lower concentrations. Peptides with anti-microbial activity, e.g. histidine-rich polypeptides, are also present in saliva. The role of these factors in controlling the resident oral microflora will be discussed in a later section.

Table 2.2 Specific and non-specific host defence factors of the mouth

Defence factor	Main function
Non-specific	
Saliva flow	Physical removal of micro-organisms
Mucin/agglutinins	Physical removal of micro-organisms
Lysozyme-protease-anion system	Cell lysis
Lactoferrin	Iron sequestration
Apo-lactoferrin	Cell killing
Sialoperoxidase system	Hypothiocyanite production (neutral pH)
	Hypocyanous acid production (low pH)
Histidine-rich peptides	Antibacterial and antifungal activity
Specific	
Intra-epithelial lymphocytes	Cellular barrier to penetrating bacteria and/or antigens
Langerhans cells	
sIga	Prevents microbial adhesion and metabolism
IgG, IgA, IgM	Prevent microbial adhesion; opsonins; complement activators
Complement	Activates neutrophils
Neutrophils/macrophages	Phagocytosis

Gingival crevicular fluid, GCF

Serum components can reach the mouth by the flow of a serum-like fluid through the junctional epithelium of the gingivae (Table 2.1). The flow of this gingival crevicular fluid (GCF) is relatively slow at healthy sites (approximately 0.3µl/tooth/hour). In periodontal disease, the production of GCF rises during inflammation although the actual flow rate may not necessarily increase because of the increase in surface area as a result of pocket formation. GCF can influence the ecology of the site in a number of ways. Its flow will remove non-adherent microbial cells, but it will also act as the primary source of nutrients for the resident micro-organisms. Many bacteria from sub-gingival plaque are proteolytic and interact synergistically to break down the host proteins and glycoproteins to provide peptides, amino acids and carbohydrates for growth. Essential co-factors, including haemin for black-pigmented anaerobes, can also be obtained from the degradation of haeme-containing molecules such as transferrin, haemopexin, haemoglobin and haptoglobin.

The increased production of gingival crevicular fluid during disease can lead to a small rise in the pH of the pocket. The mean pH during health is approximately 6.90 and this can rise during gingivitis and periodontal disease to between pH 7.25–7.75. Even such a modest change in pH could affect the proportions and types of bacteria able to grow, with the growth of some of the putative periodontal pathogens being favoured by alkaline environmental pH values. The activity of some proteases associated with the virulence of these pathogens is also enhanced at alkaline pH (pH 7.5–8.0). The most likely explanation for the rise in pH is the bacterial production of ammonia from urea and from the deamination of amino acids, the concentrations of which are elevated during inflammation.

GCF also contains components of the host defences (Tables 2.1 and 2.2) which will play an important role in regulating the microflora of the gingival crevice in health and disease. In contrast to saliva, IgG is the predominant immunoglobulin; IgM and IgA are also present, as is complement. GCF contains leucocytes, of which 95% are neutrophils, the remainder being lymphocytes and monocytes. The neutrophils in GCF are viable and therefore have the capacity to phagocytose bacteria within the crevice. A number of enzymes can be detected in GCF, e.g. collagenase, elastase, trypsin, etc, which are derived both from phagocytic host cells as well as from bacteria. These enzymes can degrade host tissues and thereby contribute to the destructive processes associated with periodontal diseases (Chapter 7). Several of these enzymes are under evaluation as candidates for diagnostic markers of active peridontal breakdown.

FACTORS AFFECTING THE GROWTH OF MICRO-
ORGANISMS IN THE ORAL CAVITY

Many factors influence the growth of micro-organisms. Some of particular relevance to microbial growth in the oral cavity will be considered in the following sections.

Temperature

Temperature is important not only because of its effect on bacterial metabolism and enzyme activity, but also due to its effect on the habitat. Parameters that can be influenced by temperature include pH, ion activity, aggregation of macromolecules and the solubility of gases. The human mouth is kept at a relatively constant temperature (35–36°C) which provides stable conditions suitable for the growth of a wide range of micro-organisms.

Redox potential/anaerobiosis

Despite the easy access to the mouth of air with an oxygen concentration of approximately 20% (155 mm Hg pO_2), it is perhaps surprising that the oral microflora comprises few, if any, truly aerobic (oxygen-requiring) species. The major organisms are either facultatively anaerobic (i.e. can grow in the presence or absence of oxygen) or obligately anaerobic (i.e. oxygen is toxic to these organisms). In addition, there are some capnophilic (CO_2-requiring) and micro-aerophilic species (requiring low concentrations of oxygen for growth). Anaerobiosis is frequently described in rigid terms: micro-organisms are separated into aerobes and anaerobes on their ability to grow in the presence or absence of oxygen. However, sharp distinctions cannot be made between these groups. A wide spectrum of oxygen tolerances occur among these organisms, including those isolated from the mouth.

Oxygen concentration is considered the main factor limiting the growth of obligately anaerobic bacteria. In the majority of microbial habitats it is the commonest and most readily reduced electron acceptor and its presence results in the oxidation of the environment. Anaerobic species require reduced conditions for their normal metabolism; therefore, it is the degree of oxidation-reduction at a site that governs the survival of these organisms. This oxidation-reduction level is usually expressed as the redox potential (Eh). Oxygen is only one of the many interacting components influencing the Eh of a habitat and its inhibitory action is usually attributed to its ability to raise the redox potential. Even if oxygen is totally excluded from the environment some anaerobes will not grow if the redox potential is too high. Similarly, some strains can survive increased concentrations of oxygen if the Eh is maintained at low levels. However, oxygen can be toxic to many anaerobes so that some oral populations may not be found in areas where these species might otherwise be expected to be present. In general, the distribution of anaerobes in the mouth will be related to the redox potential at a particular site.

Few studies have been made to determine the degree of anaerobiosis in various areas of the mouth. One study has shown that the oxygen tension of the anterior surface of the tongue is 16.4%, the posterior surface 12.4%, and the buccal folds of the upper and lower jaw only 0.3–0.4%. Micro-electrodes have enabled the redox potential to be determined at specific sites in the oral cavity. The redox potential has been shown to fall during plaque development on a clean enamel surface. The initial Eh of over +200 mV (highly oxidized) falls to as low as −141 mV (highly reduced) after seven days. The development of plaque in this way is associated with a specific succession of micro-organisms (Chapters 4 and 5). Early colonizers will utilize O_2 and produce CO_2; later colonizers may produce H_2 and other reducing agents such as sulphur-containing compounds and volatile fermentation products. Thus the Eh is gradually lowered making sites suitable for the survival and growth of a changing pattern of organisms.

Differences have been found between the Eh of the gingival crevice in health and disease. Periodontal pockets are more reduced (mean value −48 mV) than healthy gingival crevices in the same individuals (mean value +73 mV). Variations in redox potential are found between subjects; the range for the periodontal pockets in one study was +12 mV to −57 mV, and the gingival crevice +10 mV to +113 mV. Others have reported redox potentials as low as −300 mV in the gingival crevice. This is to be expected since oral spirochaetes which require an Eh in the order of −185 mV for growth can be isolated from the mouths of many individuals. Approximal areas (between teeth) will also have a low Eh although values for the redox potential at these sites have not been reported. The Eh of saliva varies considerably among individuals although it is always high. It is important to appreciate that gradients of O_2 concentration and Eh will exist in the oral cavity, particularly in plaque. Thus, plaque will be suitable for the growth of bacteria with a range of oxygen tolerances. The redox potential at various depths will be influenced by the metabolism of the organisms present and the ability of gases to diffuse in and out of plaque. Modifications to the habitat that disturb such gradients may influence the composition of the microbial community. Similarly, the metabolism or properties of particular bacteria might be influenced by the Eh of the environment since variations in RNA content and in antigenicity were found when *Actinomyces naeslundii* was grown at different redox potentials. The activity of intracellular glycolytic enzymes and the pattern of fermentation products of *Streptococcus mutans* also varies under strictly anaerobic conditions. However, more work is necessary on this topic before the full impact of redox potential and oxygen on the metabolism of oral bacteria will be understood.

pH

Many micro-organisms require a pH around neutrality for growth, and are sensitive to extremes of acid or alkali. The pH of most surfaces of the mouth is regulated by saliva. The mean pH for unstimulated whole saliva is in the range 6.75–7.25 so that, in general, optimum pH values for microbial growth will be provided at sites bathed by this fluid.

Bacterial population shifts within the plaque microflora can occur following marked fluctuations in environmental pH. After sugar consumption, the pH in plaque can fall rapidly to below pH 5.0 by the production of acids (predominantly lactic acid) by bacterial metabolism (Figure 2.3); the pH then recovers slowly to base-line values. Depending on the frequency of sugar intake, the bacteria in plaque will be exposed to varying challenges of low pH. Many of the predominant bacterial components of dental plaque associated with healthy sites can tolerate brief conditions of low pH, but are inhibited or killed by more frequent or prolonged exposures to acidic conditions. These conditions are likely to occur in subjects who commonly consume sugar-containing snacks or drinks between meals (Figure 2.4). This can result in the enhanced growth of, or colonization by, acid-tolerant (aciduric) species, such as *Streptococcus mutans* (and related species) and *Lactobacillus* species, which are normally absent or only minor components in dental plaque. Such a change in the bacterial composition of plaque predisposes a surface to dental caries.

In contrast, the pH can become more alkaline during the host inflammatory response in periodontal disease. The measurement of the pH of GCF has proved to be difficult technically. Most studies are in agreement that the pH of the healthy gingival crevice is approximately pH 6.90. Early studies using metallic antimony electrodes suggested that the pH at sites with inflammation might be as high as pH 8.50. More recently, it has been shown that such high values may be artifactual due to loss of CO_2 to the atmosphere or to the presence of reducing agents in GCF. Measurements *in situ* with glass micro-electrodes suggest that the mean pH may rise to between pH 7.2–7.4 during disease, with a few patients having pockets with a mean pH of around 7.8.

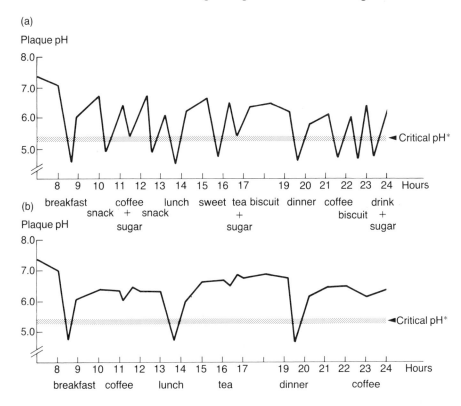

Figure 2.4 Schematic representation of the changes in plaque pH in an individual who (a) has frequent intakes of fermentable carbohydrate during the day, and (b) limits their carbohydrate intake to main meals only. *The critical pH is the pH below which demineralization of enamel is favoured.

However, even this degree of change may be sufficient to perturb the balance of the resident microflora of the gingival crevice. Laboratory studies of mixed cultures of black-pigmented Gram-negative anaerobic bacteria have shown that a rise in pH from 7.0 to 7.5 can lead to *Porphyromonas* (formerly *Bacteroides*) *gingivalis* (a major pathogen in certain periodontal diseases) increasing from less than 1% of the microbial community to predominate the culture. The pH profile for growth of some oral bacteria is shown in Figure 2.5.

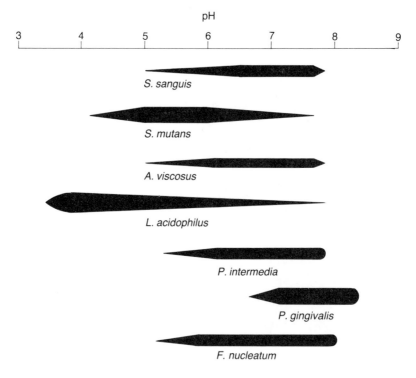

Figure 2.5 A diagrammatic representation of the pH range for growth of some oral bacteria.

Nutrients

Populations within a microbial community are dependent solely on the habitat for the nutrients essential for their growth. Therefore the association of an organism with a particular habitat is direct evidence that all of the necessary nutrients required for growth are present. In Chapter 3 it will become apparent that the mouth can support a microbial community of great diversity and satisfy the requirements of many nutritionally-demanding bacterial populations.

Endogenous nutrients

Recently, from the collective evidence of a number of studies, it has become clear that the persistence and diversity of the resident oral microflora is due to the endogenous nutrients provided by the host, rather than by exogenous factors in the diet. The main source of endogenous nutrients is saliva, which contains amino acids, peptides, proteins, vitamins, gases, and glycoproteins (which also act as a source of sugars and amino-sugars). In addition, the gingival crevice is supplied with GCF which contains albumin and other host proteins and glycoproteins, including haeme-containing molecules. The difference in source of endogenous nutrients is one of the reasons for the variation in the microflora of the gingival crevice compared with other oral sites, particularly on teeth.

Evidence for the importance of endogenous nutrients has come from the observation that a diverse microbial community persists in the mouth of humans and animals fed by intubation. Also, the oral microflora of animals with dietary habits ranging from insectivores and herbivores to carnivores is broadly similar.

Likewise, it was found that the growth rate of bacteria colonizing teeth of experimental animals was relatively unaffected by the addition of fermentable carbohydrate to their drinking water. Plaque bacteria have been found to produce glycosidases which can release carbohydrates from the oligosaccharide side chains of salivary mucins. Similarly, organisms isolated from the gingival crevice and periodontal pocket can degrade host proteins and glycoproteins including albumin, transferrin, haemoglobin, and immunoglobulins, etc. Oral micro-organisms interact synergistically to break down these endogenous nutrients. No single species has the full enzyme complement to metabolize these nutrients totally. Individual organisms possess different but overlapping patterns of enzyme activity, so that they co-operate to achieve complete degradation.

Exogenous (dietary) nutrients

Superimposed upon these endogenous nutrients is the complex array of foodstuffs ingested periodically in the diet. Despite the complexity of the diet, the only class of compounds that has been found to influence markedly the ecology of the mouth is that of fermentable carbohydrates. As discussed earlier, such carbohydrates can be broken down to acids while, additionally, sucrose can be converted into two main classes of polymer (glucans and fructans) which can be used to consolidate attachment or act as extracellular nutrient storage compounds, respectively. The frequent consumption of dietary carbohydrates is associated with a shift in the proportions of the microflora of dental plaque. The levels of mutans streptococci and lactobacilli increase, while those of acid-sensitive species such as S. sanguis decrease. The metabolism of plaque changes so that the predominant fermentation product becomes lactate. Such alterations to the microflora and its metabolism predispose a site to dental caries. Laboratory studies suggest that it is the repeated low pH generated from sugar metabolism rather than the availability of excess carbohydrate *per se* that is responsible for the perturbations to the microflora.

Dairy products (milk, cheese) may have some influence on the ecology of the mouth. The ingestion of milk or milk products can protect the teeth of animals against caries. This may be due to the buffering capacity of milk proteins or due to decarboxylation of amino acids after proteolysis since several bacterial species can metabolize casein. Milk proteins and polypeptides may also adsorb on to the tooth surface and reduce the adhesion of the resident microflora; they can also sequester calcium phosphate and enhance remineralization. Cheese has been shown to affect salivary flow rates in animals and to reduce plaque pH changes in humans following a sucrose rinse. Xylitol, a sugar substitute present in some confectionery, is not metabolized by oral bacteria. It can be inhibitory to the growth of *Streptococcus mutans*, and lower levels of this species are found in plaque and saliva of those that frequently consume confectionery containing this alternative sweetener.

Adherence

Chewing and the natural flow of saliva will detach micro-organisms not firmly attached to an oral surface. Although saliva contains between 10^8–10^9 viable micro-organisms ml^{-1}, these organisms are derived from the teeth and mucosa, with plaque and the tongue being the main contributors. Salivary components can aggregate certain bacteria which facilitates their removal from the mouth by swallowing. Bacteria are unable to maintain themselves in saliva by cell division because they are lost at an even faster rate by swallowing.

As described earlier, a unique feature of the mouth is the presence of teeth. Desquamation ensures that the bacterial load on most mucosal surfaces is light, and indeed, only a few species are able to adhere (Chapter 4). In contrast, relatively thick biofilms (dental plaque) are able to accumulate on teeth, particularly at stagnant or retentive sites like fissures, approximal regions, and the gingival crevice (Figure 2.2) which offer protection from saliva, mastication, or crevicular fluid flow. Many specific mechanisms of cell-to-host surface and cell-to-cell adherence have been determined for oral bacteria and will be described in detail in Chapter 4.

Antimicrobial agents and inhibitors

Unlike most other ecosystems, the mouth is challenged regularly with modest concentrations of antimicrobial agents and inhibitors. These are delivered mainly in the form of toothpastes (dentifrices), but the use of mouthwashes with biological activity is also on the increase. Toothpastes contain detergents such as sodium lauryl sulphate as a foaming agent. These are markedly bactericidal in laboratory tests and can lead to the reduction of salivary bacterial counts *in vivo*. Detergents are not usually retained in the mouth for long periods and so their effect is usually transitory. Fluoride is present in most toothpastes and although its primary beneficial anti-caries action is due to its incorporation into enamel and its influence on remineralization (Chapter 6), it can inhibit bacterial metabolism, particularly glycolysis, even at low concentrations under acidic conditions. In this way, fluoride may help prophylactically in the suppression of cariogenic and acid-tolerant species such as *S. mutans* under conditions in which they would otherwise flourish. Many toothpastes are now being manufactured with proven antimicrobial agents. These agents will have to be carefully monitored to ensure that their action is selective so that they inhibit organisms implicated with disease rather than health, so that the ecology of the oral microflora is not irrevocably disrupted. These inhibitors include metal ions, phenolic compounds, and plant extracts, and will be discussed in more detail in Chapters 6 and 7.

Inhibitors are also delivered via mouthrinses, and the most potent agent to date is chlorhexidine. This agent has proven antibacterial, antiviral and antifungal activity that has been confirmed in numerous clinical trials (Chapters 6 and 7). It also has marked anti-plaque activity and is widely used by handicapped individuals for whom oral hygiene is difficult. Other mouthrinses contain less active antimicrobial agents including 'essential oils' such as thymol and menthol, and plant extracts. These agents can combat oral malodours (halitosis), but also have some anti-plaque activity. Disclosing solutions contain dyes, some of which have limited antimicrobial action.

Antibiotics given systemically or orally for problems at other sites in the body will enter the mouth via saliva or GCF and affect the stability of the oral microflora. Within a few hours of taking prophylactic high doses of penicillins, the salivary microflora can be suppressed with the emergence of antibiotic-resistant bacteria. These bacteria can then persist at significant levels for several weeks before returning to their low base-line values. If several courses of treatment are necessary the antibiotic should be changed if the interval between courses is less than one month. Resistant bacteria have been observed following the use of penicillins and erythromycin. These aspects will be considered in more detail in Chapter 8.

Despite these regular environmental perturbations, the oral microflora remains relatively stable in composition and proportions. This stability is termed microbial homeostasis and the mechanisms that support this homeostasis in dental plaque are discussed in Chapter 5.

Host defences

The health of the mouth is dependent on the integrity of the mucosa (and enamel) which acts as a physical barrier to prevent penetration by micro-organisms or macromolecular antigens (Figure 2.6). The host has a number of additional defence mechanisms which play an important role in maintaining the integrity of these oral surfaces; these are listed in Table 2.2 and their sphere of influence is shown diagrammatically in Figures 2.6 and 2.7. These defences are divided into specific or immune components and non-specific, or innate, factors. The latter, unlike antibodies, do not require prior exposure to an organism or antigen for activity and so provide a continuous, broad spectrum of protection.

Non-specific factors

The physical removal by swallowing of micro-organisms by the flow of saliva (or gingival crevicular fluid) is an important defence mechanism. Bacteria can be trapped by mucins or aggregated by specific receptors on glycoproteins; these larger aggregates are more easily swallowed. Lysozyme, found in a number of bodily secretions including saliva, has the potential to hydrolyse peptidoglycan which confers rigidity to bacterial cell walls. At acid pH, the lytic action of lysozyme is enhanced by monovalent anions (bicarbonate, fluoride, chloride, or thiocyanate) and proteases found in saliva. Lactoferrin is a high affinity iron-binding glycoprotein, and since microbial pathogens grow in the host under iron-restricted conditions, the protective role of this molecule was presumed to be due primarily to iron sequestration. However, iron-free lactoferrin (apo-lactoferrin) can also be bactericidal to *Streptococcus mutans* under aerobic conditions. The salivary peroxidase enzyme system (sialoperoxidase) can generate hypothiocyanite at neutral pH or hypothiocyanous acid at low pH in the presence of H_2O_2, and both can inhibit glycolysis by plaque bacteria. H_2O_2 is generated as an end product of metabolism by several resident bacterial species, including *S. sanguis* and *S. mitis*. Recently, histidine-rich peptides, a new class of salivary antimicrobial compound, have been identified which, in the laboratory, are capable of inhibiting the growth of and killing *S. mutans* and *Candida albicans*. These peptides may help regulate the levels of yeasts in the oral cavity.

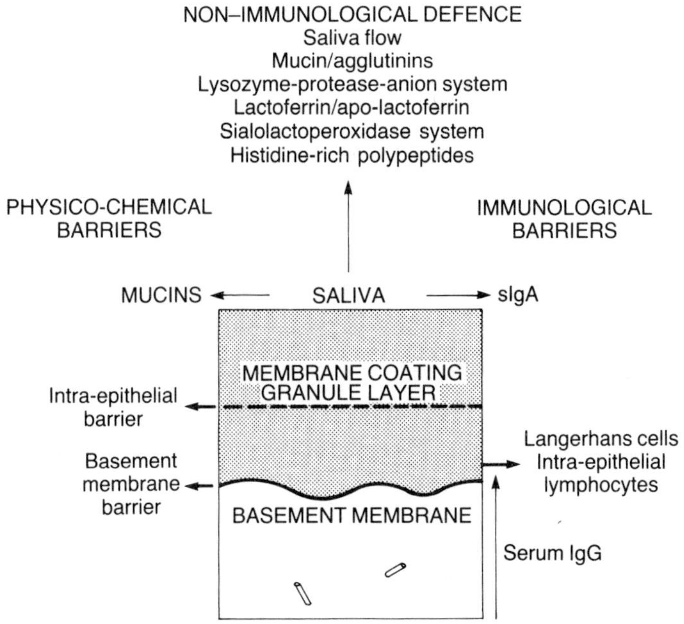

NON–IMMUNOLOGICAL DEFENCE
Saliva flow
Mucin/agglutinins
Lysozyme-protease-anion system
Lactoferrin/apo-lactoferrin
Sialolactoperoxidase system
Histidine-rich polypeptides

PHYSICO-CHEMICAL
BARRIERS

IMMUNOLOGICAL
BARRIERS

MUCINS ⟵ SALIVA ⟶ sIgA

MEMBRANE COATING
GRANULE LAYER

Intra-epithelial
barrier

Basement
membrane
barrier

Langerhans cells
Intra-epithelial
lymphocytes

BASEMENT MEMBRANE

Serum IgG

Figure 2.6 The host defences associated with oral mucosal surfaces.

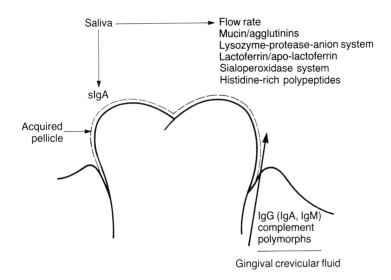

Saliva ⟶ Flow rate
Mucin/agglutinins
Lysozyme-protease-anion system
Lactoferrin/apo-lactoferrin
Sialoperoxidase system
Histidine-rich polypeptides

sIgA

Acquired
pellicle

IgG (IgA, IgM)
complement
polymorphs

Gingival crevicular fluid

Figure 2.7 Host defences associated with the tooth surfaces.

Specific factors

Components of the specific host defences (intra-epithelial lymphocytes and Langerhans cells, IgG and IgA) are found within the mucosa (Figure 2.6), where they act as a barrier to penetrating antigens. The predominant immunoglobulin in the healthy mouth is secretory IgA (sIgA). Secretory IgA can agglutinate oral bacteria, modulate enzyme activity, and inhibit the adherence of bacteria to buccal epithelium and to enamel. Compared with other classes of immunoglobulin, IgA is only weakly complement-activating and opsonizing and, therefore, is less likely to cause damage to tissues by any indirect effect of an inflammatory response. Other components, e.g. IgG, IgM, IgA, and complement, can be found in saliva but are almost entirely derived from gingival crevicular fluid (Table 2.1). GCF also contains leucocytes, of which approximately 95% are polymorphs, the remainder being lymphocytes and monocytes.

Specific antibody production can be stimulated by bacterial antigens associated with plaque at the gingival margin or on the oral mucosa. Salivary antibodies have been detected with activity against a range of oral bacteria, including streptococci, while circulating antibodies (particularly IgG) to a variety of oral microbial antigens have been reported, even in health. In the absence of inflammation, the naturally low levels of complement and polymorphs would reduce antibody-mediated phagocytosis. However, antibodies could still influence the oral microflora, either by interfering with colonization or by inhibiting metabolism. The pattern of the host response in GCF, saliva, or serum is being explored as a method of improved diagnosis of disease or as a means of recognizing at-risk individuals. Although dental diseases are widespread, the majority of disease is limited to a minority of individuals.

The antimicrobial factors described above do not necessarily operate in isolation. Combinations of specific and non-specific host defence factors can function synergistically so that, for example, lysozyme and sIgA can react with salivary agglutinins (mucins) and so be presented directly to immobilized cells. Other synergistic combinations include mucins or sIgA and sialoperoxidase. Conflicting data have been obtained as to whether the antimicrobial effect of sIgA and lactoferrin is enhanced or decreased when they are present in combination.

Despite this rich array of antimicrobial factors, the mouth naturally harbours a diverse collection of micro-organisms. Indeed, this resident microflora confers several beneficial functions on the host. The mechanisms by which the resident microflora avoids the host defences and persists at any site in the body are not fully understood. Possible strategies might include (a) antigenic variation, whereby cells undergo continuous but subtle changes in their antigenic make-up, (b) antigenic-masking, for example, by capsules, slime-layers or by the adsorption of host macromolecules onto their cell surface, (c) antigen-sharing, in which organisms either have some antigens similar to those of the host or, again, they adsorb host molecules onto their surface, and (d) inactivation of host defence factors, for example, by the production of specific proteases. Significantly, the majority of streptococci isolated from the buccal mucosa or from the early stages of plaque formation have been found to produce an IgA_1 protease. A full description of the indigenous oral microflora, together with the benefits it provides will be considered in the following chapters.

SUMMARY

The mouth is not a uniform habitat for microbial growth and colonization. Several distinct surfaces provide different habitats due to their physical nature and biological properties. These include a variety of mucosal surfaces as well as teeth; the latter are unique for microbial colonization by virtue of their being hard and non-shedding. The surfaces of the mouth are lubricated by saliva, while the gingival crevice is bathed by GCF. Both fluids help remove weakly-attached micro-organisms, and provide antimicrobial factors that help regulate bacterial and fungal colonization. Saliva and GCF are also the primary sources of nutrients for oral micro-organisms. Dietary components have much less of an influence on the microflora of the mouth, although fermentable carbohydrates can lead to increases in acidogenic and aciduric organisms that are potentially cariogenic, due to the low pH generated from their metabolism. Other factors that influence the growth of micro-organisms in the mouth include the Eh and pH of a site, the ability of cells to adhere to oral surfaces, and the presence of antimicrobial factors (from both endogenous and exogenous sources).

FURTHER READING

Arnold, R. R., Russell, J. E., Devine, S. M., Adamson, M. and Pruitt, K. M. (1985) Antimicrobial activity of the secretory innate defence factors lactoferrin, lactoperoxidase and lysozyme, in *Cariology Today* (Ed. B. Guggenheim) S. Karger, Basel, pp. 75–88.
Beighton, D., Smith, K. and Hayday, H. (1986) The growth of bacteria and the production of exoglycosidic enzymes in the dental plaque of macaque monkeys. *Archives of Oral Biology*, **31**, 829–35.

Cimasoni, G. (1983) *The Crevicular Fluid Updated*. (ed. H. M. Myers), S. Karger, Basel.

Dolby, A. E. (1986) The host defence system of the mouth. In *Immunological Aspects of Oral Diseases*. (ed. L. Ivanyi) MTP Press, Lancaster, pp. 1–11.

Dolby, A. E., Walker, D. M. and Matthews, N. (1981) *Introduction to Oral Immunology*. Edward Arnold, London.

Edgar, W. M. and O'Mullane, D. M. (eds.) (1990) *Saliva and Dental Health*. British Dental Journal, London.

Globerman, D. Y. and Kleinberg, I. (1979) Intra-oral pO_2 and its relation to bacterial accumulation on the oral tissues, in *Saliva and Dental Caries*. (eds. I. Kleinberg, S. A. Ellison and I. D. Mandel) IRL Press, London, pp. 275–91.

Igarashi, K., Lee, I. K. and Schachtele, C. F. (1989) Comparison of *in vivo* human dental plaque pH changes within artificial fissures and at interproximal sites. *Caries Research*, **23**, 417–22.

Jenkins, G. N. (1978) *The Physiology and Biochemistry of the Mouth*. 4th edn. Blackwell, Oxford.

Malamud, D. (1985) Influence of salivary proteins on the fate of oral bacteria, in *Molecular Basis of Oral Microbial Adhesion*. (eds. S. E. Mergenhagen and B. Rosan) ASM, Washington DC, pp. 117–24.

Mandel, I. D. and Ellison, S. A. (1985) The biological significance of the non-immunological defence factors, in *The Lactoperoxidase System: Chemistry and Biological Significance*. (eds. K. M. Pruitt and J. Tenovuo) Marcel Dekker, New York, pp. 1–14.

Roitt, I. M. and Lehner, T. (1983) *Immunology of Oral Diseases*. 2nd edn. Blackwell, Oxford.

Schachtele, C. F. and Jensen, M. E. (1982) Comparison of methods for monitoring changes in the pH of human dental plaque. *Journal of Dental Research*, **61**, 1117–25.

3 The resident oral microflora

The resident oral microflora is diverse and consists of a wide range of species of viruses, bacteria, yeasts and even on occasions, protozoa. This diversity is due to the fact that the mouth is composed of a number of varied habitats supplied with a variety of nutrients. In addition, in dental plaque, gradients develop in parameters of ecological significance, such as oxygen tension and pH, providing conditions suitable for the growth and survival of micro-organisms with a wide spectrum of requirements. Under such conditions, no one bacterial population has a particular advantage and numerous species can co-exist. This situation in the mouth can be contrasted with ecosystems in which the intensity of one or more parameters approaches the extreme. The microbial community from such habitats is inevitably characterized by a low species diversity.

Before the microbial community at individual sites in the mouth can be considered in detail (Chapters 4 and 5), the types and properties of the organisms found commonly in health and disease will be described. First, however, it may be instructive to discuss the principles of microbial classification (or taxonomy) and describe briefly some of the methods used.

PRINCIPLES OF MICROBIAL TAXONOMY

The purpose of taxonomy is to develop a logical arrangement of organisms based on their natural relationships. This requires the determination of as many characterstics as possible, with organisms arranged into groups of high mutual similarity. As the properties of an organism are dictated by what is coded in its genome, the similarity in DNA base composition between strains can be compared. The parameter most often used is the mole percentage of guanine (G) plus cytosine (C) in the total DNA. The % G+C content of a cell may be measured directly or calculated indirectly from buoyant density or melting point determinations. Organisms with markedly different G+C contents are unrelated, while organisms that are closely related have similar G+C values. Similarity in gross DNA composition is not necessarily proof of close relatedness, however, because the base pairs could be organized in a different sequence. In such a situation genotypic similarity can be confirmed by determining the abilities of heat-denatured single strands of DNA from different strains to re-anneal with each other during slow cooling. High levels of homology reflect an overall similarity in the nucleotide sequences from the DNA of the two strains being compared, and hence confirm the close taxonomic relationship of the strains. Generally, strains within a species share more than 70% sequence homology.

Improvements in classification have also come from the application of chemical analyses for the comparison of microbial properties (chemotaxonomy). These include the analysis of membrane lipids, peptidoglycan structure, enzyme mobilities and whole-cell protein patterns derived from polyacrylamide gel electrophoresis.

PRINCIPLES OF MICROBIAL IDENTIFICATION

Once organisms have been correctly classified using rigorous techniques, then more simple identification schemes can be devised in which limited numbers of key discriminatory properties are used. These might include sugar fermentation patterns, polysaccharide production, gas or ammonia production, or the pattern of acidic fermentation products following glucose metabolism. The type of tests will vary with the particular groups (taxa) of micro-organisms being examined. More recently, the rapid detection (four hours) of preformed enzymes in concentrated suspensions of bacteria using colorimetric and fluorimetric substrates has speeded up and improved the identification of many bacteria. The substrates for many of these enzymes have been incorporated into kit form and sold commercially. Patterns of substrates hydrolysed can then be compared with profiles held in a computerized data base and identification achieved. Furthermore, monoclonal antibodies and DNA (gene) probes have also been developed for the rapid detection and identification of a limited number of species, but primarily those associated with disease. These techniques have the advantage that organisms can be detected directly in plaque samples without the need for lengthy culturing. Some companies have been formed on the basis of offering a service of identifying periodontal pathogens in subgingival plaque from patients.

In general, the first stage in identification is to determine the reaction of an organism in the Gram stain and to ascertain its cellular morphology. Some distinguishing parameters used in microbial identification and taxonomy are listed in Table 3.1. Advances have been made recently following the use of novel, but ecologically-relevant substrates, such as those for detecting enzymes (glycosidases) that cleave sugar residues from salivary mucins. Their use has enabled species to be separated using simple tests where previously they had been difficult to distinguish from one another, e.g. some oral streptococci.

Table 3.1 Some characteristics used in microbial classification and identification schemes

Characteristic	Examples
Cellular morphology	Shape; Gram stain reaction; flagella; spores; size
Colonial appearance	Pigment; haemolysis; shape
Carbohydrate fermentation	Acid or gas production
Amino acid hydrolysis	Ammonia production
Pattern of fermentation products	Butyrate; lactate; acetate
Preformed enzymes	Glycosidases
Antigens	Monoclonal/polyclonal antibodies to cell surface proteins
Lipids	Menaquinones; long-chain fatty acids
DNA	Base composition; base sequence homology
Enzyme profile	Malate dehydrogenase
Peptidoglycan	Amino acid composition

RECENT ADVANCES IN MICROBIAL TAXONOMY

The problems

Although recent advances have led to improvements in the classification of oral bacteria, such improvements have also generated a number of difficulties when interpreting or comparing early data when a previous nomenclature was in use. The taxonomy of many groups of oral bacteria has been revolutionized in a relatively short time period, with many new genera and species described. An organism discussed in an early study may now have been reclassified and hence renamed. As it is not always clear how such an organism should be classified in modern schemes, new and old terminologies may co-exist in the text. For example, *Streptococcus sanguis* has been reported in the literature for many decades but, as of 1989, its description is now more limited and organisms that were previously included within this species are now known to be sufficiently different as to warrant a distinct species epithet, e.g. *S. gordonii*. Some strains reported in earlier studies as *S. sanguis* may not have the same properties as strains more recently identified as *S. sanguis sensu stricto*. Thus, care has to be taken when interpreting older (and not so old) scientific literature. Because of these rapid changes in nomenclature, a number of lists of synonyms for particular bacteria are given in this chapter.

Microbial taxonomy is a dynamic area with existing species being reclassified due to the application of more stringent tests and more recent techniques, together with the recognition of genuinely newly-discovered species from sites such as periodontal pockets. The emphasis paid to the taxonomy and identification of the oral microflora is necessary because without valid sub-division and accurate identification of isolates, the specific association of species with particular diseases (microbial aetiology) cannot be recognized. Likewise, it has to be accepted that further changes will occur in the future. The properties of the main groups of micro-organism found in the mouth will now be discussed.

GRAM-POSITIVE COCCI

Streptococcus

Streptococci have been isolated from all sites in the mouth and comprise a large proportion of the resident oral microflora. On average, streptococci represent 28% of the total cultivable microflora from supragingival dental plaque, 29% from the gingival crevice, 45% from the tongue, and 46% from saliva. The majority are alpha-haemolytic on blood agar and early workers grouped them together, calling them viridans streptococci. However, haemolysis is not a particularly reliable property in distinguishing these streptococci, and many oral species contain strains showing all three types of haemolysis (alpha, beta and gamma). In addition, recent taxonomic studies have shown that these 'viridans'-group streptococci can be divided into many well-defined species. Thus, where possible the term viridans streptococcus (or *Streptococcus viridans*) should not be used as it is a generalization for several species each with distinct properties.

Traditionally, oral streptococci have been differentiated by simple biochemical and physiological tests. Recent studies comparing DNA homology, whole cell-derived polypeptide patterns, and the detection of glycosidase activity has clarified the taxonomic relationship between many species, and allowed the sub-division of taxa that were acknowledged to be heterogeneous. These studies also led to the creation of two new genera, *Enterococcus* and *Lactococcus*, to include the former 'faecal' and 'dairy' groups of streptococci, respectively.

For convenience, the oral streptococci will be discussed in the following sections on the basis of their division into four main species-groups: *S. mutans*-group; *S. salivarius*-group; *S. milleri*-group and the *S. oralis*-group (Table 3.2).

Table 3.2 Currently-recognized species of oral streptococci

Group	Species
S. mutans-group (mutans streptococci)	*S. mutans* *S. sobrinus* *S. cricetus* *S. rattus* *S. ferus* *S. macacae* *S. downei*
S. salivarius-group	*S. salivarius* *S. vestibularis*
S. milleri-group	*S. constellatus* *S. intermedius* *S. anginosus*
S. oralis-group	*S. sanguis* *S. gordonii* *S. parasanguis* *S. oralis* *S. mitis* (tufted fibril group, *S. crista*)

Streptococcus mutans-group (mutans streptococci)

More literature has been published on this group of bacteria than any other oral organism because of their role in the microbial aetiology of dental caries. *S. mutans* was originally isolated from carious human teeth by Clarke in 1924, and, shortly afterwards, it was also recovered from a case of bacterial endocarditis. Little attention was paid to this species until the 1960s when it was demonstrated that caries could be experimentally-induced and transmitted in animals artificially-infected with strains resembling *S. mutans*. Its name derives from the fact that cells can lose their coccal morphology and often appear as short rods or as cocco-bacilli. Eight serotypes (serovars) were eventually recognized (*a–h*), based on the serological specificity of carbohydrate antigens located in the cell wall. Subsequent work showed that differences existed between clusters of these serotypes in a number of simple physiological characteristics (e.g. fermentation of substrates such as raffinose, melibiose and inulin, in H_2O_2 production, and in their sensitivity to bacitracin) as well as in the electrophoretic mobility of enzymes, in their profile of cell wall proteins, in peptidoglycan type, and in DNA base composition. As a result of this genetic and phenotypic heterogeneity, seven distinct species of human and animal mutans streptococci have been described, and these are listed in Table 3.3.

Table 3.3 Classification of the mutans streptococci group

Species	Previous nomenclature	Host
S. mutans	S. mutans serotypes c, e or f	Human
S. sobrinus	S. mutans serotypes d or g	Human
S. cricetus	S. mutans serotype a	Human
S. ferus*†		Rat
S. rattus	S. mutans serotype b	Human/Rodent
S. macacae*		Monkey
S. downei	S. mutans serotype h	Monkey

* *S. ferus* and *S. macacae* have the same rhamnose/glucose cell wall polysaccharide antigen as human *S. mutans* strains

† The taxonomic position of *S. ferus* is less certain as it shares many similarities with those of the *S. oralis*-group. Unlike the other mutans streptococci, *S. ferus* is also non-cariogenic in animal models.

The term *S. mutans* is now limited to human isolates previously belonging to serotypes *c, e* and *f*. This is the species of mutans streptococci most commonly isolated from dental plaque, and epidemiological studies have implicated *S. mutans* as the primary pathogen in the aetiology of enamel caries in children and young adults, root surface caries in the elderly, and nursing (or bottle) caries in infants (Chapter 6). The next most commonly-isolated species of the mutans streptococci group is *S. sobrinus*, which has also been associated with human dental caries. However, there is less known about the role of *S. sobrinus* in disease than *S. mutans* because some commonly-used selective media for the isolation of mutans streptococci from dental plaque contain bacitracin which can be inhibitory to the growth of both *S. sobrinus* and *S. cricetus*. Studies on the preferred oral habitat of the most prevalent mutans streptococci have suggested that the presence of *S. mutans* is similar on all teeth, but it is found more often and in higher numbers than *S. sobrinus* in fissures (Figure 2.2), while *S. sobrinus* is isolated preferentially from posterior rather than from anterior teeth.

Mutans streptococci are isolated regularly from dental plaque, but their prevalence is low on sound enamel. The isolation frequency of *S. mutans* is high, irrespective of age, race, or geographical location of the subjects studied (Table 3.4). *S. sobrinus* is also commonly isolated but usually at a lower frequency than *S. mutans*, while *S. cricetus* and *S. rattus* are recovered only rarely. Some studies have reported that *S. sobrinus* is found more commonly on denture-plaque than *S. mutans*, but other studies have failed to confirm this observation. Some subjects harbour more than one species of mutans streptococci in their mouth. These range from *S. mutans/S. sobrinus* combinations which are found together in 3% (UK children) to 49% (Saudi Arabian adults) of individuals, to combinations of *S. mutans* and *S. rattus*, and *S. mutans* and *S. cricetus*, which were found together in 1% of elderly individuals in the USA.

The antigenic structure of mutans streptococci has been studied in detail because of its possible application in anti-caries vaccine development (Chapter 6) or in serological typing schemes. Mutans streptococci possess cell wall carbohydrate antigens, lipoteichoic acid (which extends from the cytoplasmic membrane) and cell wall or cell wall-associated proteins. The proteins have generated considerable interest because of their role as immunogens in a sub-unit or synthetic anti-caries vaccine. The antigen that has been studied most intensively has been termed I/II, antigen B, or antigen P1 by various authors, and is a high molecular weight protein, although the immunologically-reactive epitope can be retained in smaller fragments. This antigen may be involved in the initial adherence of *S. mutans* to the tooth surface by interacting with components of the salivary pellicle (Chapters 4 and 5). A similar protein is designated SpaA in *S. sobrinus*. Many of the protein antigens of mutans streptococci are extracellular and can be recovered and purified from culture supernatants.

Differentiation of mutans streptococci into the species listed in Table 3.3 can be problematical. Some physiological distinctions do exist, but they are not always reliable. As an alternative approach, specific monoclonal or polyclonal antibodies have been raised to identify individual species either by immunofluorescence or by immunoblotting techniques of colonies.

Table 3.4 Prevalence* of mutans streptococci in different population groups

Population Group (Country)	Mutans streptococci			
	S. mutans	S. sobrinus	S. cricetus	S. rattus
Infants (USA)	74	22	0	0
Infants (Japan)	82	36	0	0
Children (USA)	95	<1	0	<1
Children (UK)	70	7	–	–
Children (Canada)	97	3	0	0
Children (Japan)	>90	30	0	0
Children (Tanzania)	0	17	0	20
Children (Columbia)	>65	25	0	0
Children (China)	58	41	0	<1
Children (Iceland)	>76	35	–	–
Adults (USA)	>90	5–10	0–16	0
Adults (Netherlands)	80–100	5–40	0	0
Adults (Europe)	56	52	3	8
Adults (Saudi Arabia)	76	53	0	0
Adults (Mozambique)	81	26	–	83
Elderly (Sweden)	72	44	–	–
Elderly (USA)	>85	7	2	3

* Prevalence is expressed as the percentage of a population group in which the particular species was detected.

Mutans streptococci make soluble and insoluble extracellular polysaccharides from sucrose that are associated with plaque formation and cariogenicity (Chapters 3 and 4). They can also synthesize intracellular polysaccharides which act as carbohydrate reserves, and can be converted to acid during periods when dietary carbohydrates are unavailable. Mutans streptococci produce acid at an extremely rapid rate from pulses of fermentable carbohydrate, and this no doubt contributes to their pathogenicity in caries. However, of equal significance is their ability to grow and survive under the acidic conditions they generate.

Steptococcus salivarius-group

The S. salivarius group comprises S. salivarius, the closely-related S. thermophilus (which is not found in the mouth) and the newly-described

S. vestibularis. Strains of *S. salivarius* are commonly isolated from most areas of the mouth although they preferentially colonize mucosal surfaces, especially the tongue. They produce an extracellular levan (polymer of fructose) from sucrose (Chapter 4), and this gives rise to characteristically-large mucoid colonies when grown on sucrose-containing agar. This polymer is highly labile and may be metabolized in the mouth by other oral organisms (Chapter 5). *S. salivarius* is isolated only occasionally from diseased sites, and is not considered a significant opportunistic pathogen.

Recently, a new species, *S. vestibularis*, was recognized consisting of α-haemolytic streptococci isolated mainly from the vestibular mucosa of the human mouth. These strains are unable to produce extracellular polysaccharides from sucrose, but do produce a urease (which can generate ammonia and hence raise the local pH) and hydrogen peroxide (which can contribute to the sialoperoxidase system, and inhibit the growth of competing bacteria). There is little known about the ecology of this species at present. *S. thermophilus* is not recovered from the human mouth but it can be isolated from milk and dairy products. Previously, it was described as *S. salivarius* subsp. *thermophilus*.

Streptococcus milleri-group

Organisms within this group are isolated readily from dental plaque and from mucosal surfaces, but they are also an important cause of purulent disease in humans. In one study they were the most common streptococci found in abscesses of internal organs, especially of the brain and liver, and they have also been recovered from cases of appendicitis, peritonitis and endocarditis. The classification and nomenclature of this group has been confused for a number of years, although recent chemotaxonomic studies comparing DNA sequence homologies, whole cell polypeptide patterns and other phenotypic traits have resolved the situation. Early work in Europe had grouped various haemolytic and non-haemolytic streptococci, including those with Lancefield group F and G antigens, as *S. milleri*. In the USA, an alternative classification had been proposed and two groups were formed, designated *S. MG-intermedius* and *S. anginosus-constellatus*. To resolve the situation, it was subsequently proposed that strains should be divided on the basis of lactose fermentation into *S. constellatus* (non-lactose fermenters) and *S. intermedius* (lactose fermenters), with *S. anginosus* to include a collection of β-haemolytic strains possessing various Lancefield grouping antigens. The latest chemotaxonomic studies confirm the existence of these three distinct species, although their precise descriptions have been emended. In future, the term *S. milleri* will cease to be used. Strains from this group of bacteria have been known for many years to be potential opportunistic pathogens. From a clinical viewpoint, *S. interme-dius* appears to be isolated mainly from liver and brain abscesses and most strains produce a hyaluronidase, while *S. anginosus* and *S. constella-tus* are derived from a wider range of sources. No strains from this group make extracellular polysaccharides from sucrose.

Streptococcus oralis-group

This group has also been acknowledged for many years to be hetero-geneous, and has suffered many changes in nomenclature. As with the *S. milleri*-group, alternative nomenclatures have been in common use. Again, the recent application of chemotaxonomic techniques has resolved many of the differences and anomalies. *S. sanguis* produces extracellular soluble and insoluble glucans (polymers of glucose, Chap-ter 4) from sucrose, which are implicated in the development of plaque. Originally, *S. sanguis* was differentiated into two groups: *S. sanguis* I and II. *S. sanguis* I has now been further separated into two distinct species: *S. sanguis* and *S. gordonii*. The former produces a protease that can cleave IgA, while the latter can bind α-amylase which can remain active in this state and break down starch (Table 3.5). Amylase-binding may also mask bacterial antigens and allow the organism to avoid

Table 3.5 Past and present nomenclature of streptococci belonging to the *S. oralis*-group

Previous nomenclature	Present nomenclature	Distinguishing properties
S. sanguis I	S. sanguis	IgA protease production
	S. gordonii	α-amylase binding
S. sanguis II/'S. mitior'/S.mitis	S. oralis	neuraminidase production IgA protease production glucan production
	S. mitis	α-amylase binding noglucan production

Other species within this group are *S. parasanguis* and the tufted fibril strains.

recognition by the host defences. Even with this new classification, three biotypes have been recognized within *S. gordonii* and four biotypes within *S. sanguis*.

Strains originally designated *S. sanguis* II were termed *S. mitior* by most European workers. Some strains produced glucans (dextran) from sucrose while others did not; these latter strains were designated by some as *S. mitis*. Unfortunately, the widely-used name *S. mitior* is not officially recognized and should be replaced by the internationally valid name *S. oralis*. Strains of *S. oralis* produce neuraminidase and an IgA₁ protease but cannot bind α-amylase. Extracellular glucan production from sucrose is a variable characteristic. A closer examination of strains designated as *S. mitis* has identified two biotypes: biotype 1 strains carry the Lancefield group O antigen and cannot hydrolyse arginine to ammonia while some strains of biotype 2 carry the Lancefield group K antigen and all can hydrolyse arginine. Some strains of *S. mitis* produce an IgA₁ protease, especially those found in the early stages of plaque formation or colonizing the buccal mucosa. Some strains also have neuraminidase activity but, in the current description of the species, none make glucan or fructan from sucrose.

The revised classification (Table 3.5) has allowed more specific eco-logical relationships between bacterium and host to be recognized. For example, in the early stages of plaque development, the likely preva-lence of species will be:

$$S.\ mitis\ 1 > S.\ oralis > S.\ sanguis > S.\ mitis\ 2 > S.\ gordonii$$

As this nomenclature gets more widely used, then other trends may emerge.

Members of the *S. oralis*-group have long been recognized as potential opportunistic pathogens. For example, infective endocarditis results from the growth of blood-borne bacteria on damaged heart valves (Chapter 8). Members of the *S. oralis*-group are isolated most frequently from this serious condition, and are assumed to have originated from the mouth; mutans streptococci are also isolated commonly from these patients.

Recently, the name *S. parasanguis* has been proposed for a group of strains isolated from clinical specimens (throat, blood, urine). Strains can hydrolyse arginine but not urea, and can bind salivary α-amylase, but cannot produce extracellular polysaccharides from sucrose. The ecology and habitat of this species is not certain at present.

Another sub-group of strains has been recognized that resemble *S. sanguis* but are characterized by the presence of tufts of fibrils on their cell surface (Chapter 4). The name *S. crista* has been proposed for this group of organisms.

Other streptococci

There is little definitive information on the isolation and prevalence of obligately anaerobic Gram-positive cocci from the mouth. Confusion surrounds their classification, while considerable difficulties can be encountered in their isolation from oral samples. Two genera were originally proposed: *Peptostreptococcus*, which grow in chains and *Peptococcus*, which grow in clumps. However, most species of the latter have now been transferred to *Peptostreptococcus* and several peptostreptococcal species have been recovered from dental plaque (e.g. *Ps. micros, Ps. magnus* and *Ps. anaerobius*), especially from advanced forms of periodontal disease (Chapter 7), carious dentine, infected pulp chambers and root canals (Chapter 6), and dental abscesses (Chapter 8). Any potential pathogenic role for these organisms in such conditions is unclear at present.

Lancefield group A streptococci (*S. pyogenes*) are not usually isolated from the mouth of healthy individuals, although they can often be cultured from the saliva of people suffering from streptococcal sore throats, and may be associated with a particularly acute form of gingivitis (Chapter 7).

Enterococcus

Little information is available on the presence of enterococci in the healthy mouth. They have been recovered in low numbers from several oral sites when appropriate selective media were used; the most frequently isolated species is *Enterococcus* (formerly *Streptococcus*) *faecalis*. Some strains of enterococci can induce dental caries in gnotobiotic rats while others have been isolated from infected root canals and from periodontal pockets, particularly in immunocompromised subjects. While *E. bovis* has not been isolated with certainty from the human mouth, it has been found in dental plaque from a range of herbivore animals.

Staphylococcus, Micrococcus and *Stomatococcus*

Staphylococci and micrococci are not commonly isolated in large numbers from the oral cavity although the former have been reported in plaque samples from subjects with dentures or with certain periodontal diseases, and in immunocompromised patients and individuals suffering from a variety of oral infections (Chapter 8). These bacteria are not usually considered to be members of the resident oral microflora, but may be present transiently.

Interestingly, this is in sharp contrast to surfaces of the human body in close proximity to the mouth, such as the skin surface and the mucous membranes of the nose. This finding emphasizes the major differences that must exist in the ecology of these particular habitats. Skin and nasal flora must be passed consistently into the mouth and yet these organisms are normally unable to colonize or compete against the resident oral microflora.

Stomatococcus (formerly *Micrococcus*) *mucilagenosus* is a catalase positive, Gram-positive coccus that is isolated almost exclusively from the tongue. It produces an extracellular slime which may play a significant role in the association of this species with that site.

GRAM-POSITIVE RODS AND FILAMENTS

Gram-positive rods and filaments are commonly isolated from dental plaque and consist of facultatively- and obligately-anaerobic species. Chemotaxonomic methods have always played a fundamental role in the identification and classification of these bacteria; the pattern of acids produced from the fermentation of glucose has proved of particular value. The properties of some of the more prevalent oral Gram-positive rods will be described in the following sections.

Actinomyces

Actinomyces species form a major portion of the microflora of dental plaque, particularly at approximal sites. These organisms also colonize the gingival crevice and may influence the subsequent pattern of plaque formation (Chapter 4). They are known to increase in numbers during gingivitis and have been associated with root surface caries (Chapters 7 and 6, respectively). Physiological and antigenic heterogeneity has been reported among strains resembling *A. viscosus* and *A. naeslundii*, and two new human species have recently been described (Table 3.6).

Strains of *A. viscosus* characteristically produce an extracellular slime and fructans from sucrose. Two serotypes of this species were described originally; strains of serotype I being associated with animals and serotype II with humans. *A. viscosus* serotype II is phenotypically and genotypically very similar to human strains of *A. naeslundii*, and were distinguished traditionally only on the basis of catalase production by the former. It has been concluded that the division of *A. viscosus* from *A. naeslundii* is closer to a serotypic separation than a separation at a species level. Strains have since been found that are variant in the catalase reaction and many now combine such organisms into a single

Table 3.6 Classification of human *Actinomyces* species

Present nomenclature	Previous designation
Actinomyces georgiae	untypable subgingival *Actinomyces* sp.
Actinomyces gerencseriae	*A. israelii* serotype II
A. israelii	*A. israelii* serotype I
A. odontolyticus	*A. odontolyticus*
A. naeslundii genospecies 1	*A. naeslundii* serotype I
A. naeslundii genospecies 2	*A. naeslundii* serotypes II and III
	A. viscosus serotype II
A. meyeri	*A. meyeri*

A. viscosus/naeslundii group, although serological distinctions can still be discerned. Particular biotypes vary in their colonization of the dentition, and may be associated with enamel and root surface caries.

Actinomyces israelii can act as an opportunistic pathogen causing a chronic inflammatory condition called actinomycosis (Chapter 8). The disease is usually associated with the cervicofacial region, but it can disseminate to cause deep-seated infections in other sites in the body such as the abdomen. Strains of *A. israelii* characteristically form 'granules' which may contribute to the ability of these bacteria to disseminate around the body by affording them physical protection from the environment, from the host defences, and from antibiotic treatment. *Actinomyces* species have also been found in cervical smears of women using intrauterine contraceptive devices. The isolation rate of these species varies according to the type of sample taken and whether organisms were identified by cultural techniques or by serological methods such as immunofluorescence, but values ranging from 15–65% have been reported. Many of the isolates have been identified as *A. israelii*, although other species including *Arachnia propionica* (now reclassified as *Propionibacterium propionicus*) have also been found. *A. israelii* has not been shown to produce any toxins although cells may be immunosuppressive and modulate the host defences at a site. This species has a complex antigenic composition comprising both carbohydrate and polypeptide antigens. Recently, *A. israelii* serotype II has been designated as a separate species, *A. gerencseriae*, which is a common but minor component of the microflora of the healthy gingival crevice, although it has also been isolated from abscesses. *A. israelii* and *A. gerencseriae* are both obligately anaerobic but can be differentiated on the basis of whole cell protein banding patterns, serological reactions, and the inability of the latter to ferment L-arabinose.

Another newly described species is *A. georgiae*, which is a facultatively anaerobic organism that was found occasionally in the healthy gingival microflora. Other species include *A. odontolyticus*, of which about 50% of strains form colonies with a characteristic red-brown pigment. Two serotypes have been reported and recent studies have found an association between this species and the very earliest stages of enamel demineralization, and with the progression of small caries lesions (Chapter 6). The species *A. meyeri* has been reinstated; this species has been reported occasionally and in low numbers from the gingival crevice in health and disease, and from brain abscesses.

Distinct *Actinomyces* species from animals have been described. These include *A. hordeovulneris* from dogs and *A. bovis* from cattle and a llama. Several other species have also been recognized in dairy cattle and include *A. slackii* (catalase positive), *A. denticolens*, and *A. howellii* (both catalase negative). *A. pyogenes* is found on the mucosal surfaces of warm-blooded animals and is reported to be pathogenic for man.

Eubacterium

This is a poorly-defined genus which contains a variety of obligately anaerobic, filamentous bacteria that are often Gram-variable. Traditionally, this genus has been regarded for some time as a 'dumping-ground' for strains not belonging to other genera of anaerobic, non-spore-forming bacilli. They can be difficult to cultivate and many laboratories have not isolated them from plaque while others have found up to 17 'taxa' in various forms of periodontal disease. The classification and identification of these strains is hampered severely because of the general lack of reactivity of isolates. Consequently, whole cell protein profiles, patterns of fermentation products, and the presence of preformed enzymes have to be used in an attempt to differentiate between species. Among the proposed species are *E. brachy*, *E. timidum* and *E. nodatum*; strains are asaccharolytic and found in subgingival plaque, particularly during periodontal disease. *E. yurii* has also been isolated from subgingival plaque and is involved in some of the 'corn-cob' and 'test-tube brush' arrangements with coccal bacteria seen in dental plaque (Chapter 5). Other species that have been described include *E. lentum*, *E. saburreum* and *E. alactolyticum*, but it is not clear whether all of these are found in the mouth. This genus requires further study especially as these organisms can be frequently isolated from numerous infections of the head, neck and lung, and from necrotic dental pulp. Interestingly, some distinct eubacteria have been isolated from animals and include

E. fossor, which has been isolated from the pharynx and from tooth abscesses of horses, and *E. suis* from pigs.

Lactobacillus

Lactobacilli are commonly isolated from the oral cavity although they usually comprise less than 1% of the total cultivable microflora. However, their proportions and prevalence increase at advanced caries lesions both of the enamel and of the root surface. A number of homo- and hetero-fermentative species have been identified, producing either lactate or lactate and acetate, respectively, from glucose. The most common species are *L. casei*, *L. fermentum* and *L. acidophilus*, with *L. salivarius*, *L. plantarum*, '*L. brevis*', *L. cellobiosus* and *L. buchneri* also reported. Recently, some of these species have been shown to be heterogeneous. For example, *L. acidophilus* has been subdivided into *L. acidophilus sensu stricto*, *L. crispatus* and *L. gasseri*, while within the *L. casei*-group, several distinct species have been recognized including *L. rhamnosus* and *L. casei* (formerly proposed as *L. paracasei*). Whether all of these species are found in the mouth has yet to be determined, although preliminary studies suggest that the majority of oral isolates are *L. rhamnosus* and *L. casei*. Comparisons of the properties of *L. brevis*-like strains from the mouth and from cheese showed that the oral strains were distinct, and these have now been placed in a new species, *L. oris*. *L. uli* has been isolated from periodontal pockets. Some *L. acidophilus*-like strains were isolated from animals (monkey, baboon, dog) and have been re-classified as *L. animalis*.

Despite these advances in classification, little is known of the preferred habitat and niche for these species in the normal mouth, and most studies still merely identify them as 'lactobacilli' or *Lactobacillus* spp. They are highly acidogenic organisms, and early work implicated lactobacilli in the initiation of dental caries. Subsequent research has shown that they are associated more with carious dentine and the advancing front of caries lesions than with the initiation of the disease. Crude tests with selective media have been designed for estimating the numbers of lactobacilli in patients' saliva to give an indication of the cariogenic potential of a mouth. Although these tests are often unreliable, they are useful for monitoring the dietary behaviour of a patient because levels of lactobacilli correlate well with the intake of dietary carbohydrate.

Propionibacterium

Several species of propionibacteria have been reported from the mouth, including *P. acnes* in dental plaque. These bacteria are obligately anaerobic and this genus now includes *P. propionicus* which was formerly classified as *Arachnia propionica*. *P. propionicus* is morphologically indistinguishable from *A. israelii* but the two species can be distinguished by the production of propionic acid from glucose by the former. Strains of *P. propionicus* have been isolated from cases of actinomycosis and lacrimal canaliculitis (infection of the tear duct), and so are opportunistic pathogens.

Other genera

Corynebacterium (formerly *Bacterionema*) *matruchotii*, *Rothia dentocariosa*, and *Bifidobacterium dentium* have been regularly isolated from dental plaque. *C. matruchotii* has an unusual cellular morphology having a long filament growing out of a short, fat rod-like cell, thus earning its description of 'whip-handle' cell. This species is found only in the mouth and appears to be the only true coryneform in the oral cavity. Occasionally strains of mycobacteria, *Corynebacterium xerosis* and some *Bacillus* and *Clostridium* species are isolated, but these probably represent merely transient visitors from other habitats.

GRAM-NEGATIVE COCCI

Neisseria

Neisseria are isolated in low numbers from most sites in the oral cavity. Along with the *S. oralis*-group they are among the early colonizers of a clean tooth surface. *Neisseria* spp. are generally saccharolytic and can grow well aerobically, although their growth is stimulated by carbon dioxide and retarded under anaerobic conditions (Table 3.7). Care has to be taken when culturing these organisms because they are autolytic. The taxonomy of this group remains confused. The most common species is *N. subflava* which is saccharolytic and polysaccharide producing; its colonies also produce a green/yellow pigment. It is rarely associated with disease. *N. sicca* is closely-related and also produces polysaccharides and is saccharolytic, but its colonies have no pigment. *N. mucosa* is found in the nasopharynx, and some strains possess a capsule. Some neisseriae are asaccharolytic and non-polysaccharide forming, such as '*N. cinerea*' which may be a commensal of the oro-pharynx. Some distinct species have been isolated from animals and include *N. canis* and *N. animalis*.

Table 3.7 Properties of human oral Gram-negative cocci

Genus	Species	Preferred atmosphere for growth	Extracellular polysaccharide production	Saccharolytic
Neisseria	subflava	aerobic	+	+
	mucosa	aerobic	C*	+
	sicca	aerobic	+	+
Branhamella	catharrhalis	aerobic	−	−
Veillonella	parvula	anaerobic	−	−
	dispar	anaerobic	−	−
	atypica	anaerobic	−	−

* Some strains produce a capsule.

In general, aerobic isolates that are asaccharolytic, non-polysaccharide formers, and whose colonies lack pigment, fall within the genus *Branhamella* (Table 3.7). *B. catharrhalis* is a commensal of the upper respiratory tract but is also a well-established opportunistic pathogen. Many strains of *B. catharrhalis* produce a β-lactamase which can complicate antibiotic treatment.

The significance in oral ecology of polysaccharide production (some of which are glycogen-like) by strains of *Neisseria* is unclear, although some streptococcal strains can metabolize these polymers (Chapter 4). Some strains of *Neisseria* also metabolize lactic acid, and the significance of this property is discussed in the next section on *Veillonella*.

Veillonella

Veillonella are strictly anaerobic Gram-negative cocci that were originally separated into two species, *V. parvula* and *V. alcalescens*, on the basis of the breakdown of hydrogen peroxide by the latter. The mechanism of breakdown is unclear because *V. alcalescens* does not produce catalase. These two species have now been combined into *V. parvula*, and two other human species recognized, *V. dispar* and *V. atypica* (Table 3.7). A number of related but distinct species have been isolated from the mouth of a wide range of rodents e.g. *V. ratti*, *V. rodentium*, *V. criceti* and *V. caviae*. Veillonellae have been isolated from most surfaces of the oral cavity although they occur in highest numbers in dental plaque. Veillonellae lack glucokinase and fructokinase and are unable to metabolize carbohydrates. They utilize several intermediary metabolites, in particular lactate, as energy sources and consequently may play an important role in the ecology of dental plaque and in the aetiology of dental caries. Apart from small quantities of formic acid, lactic acid is the strongest acid produced by oral bacteria and, therefore, it is implicated in the dissolution of enamel. It has been proposed that *Veillonella* in plaque might reduce the harmful effect of potentially cariogenic bacteria by metabolizing lactic acid, converting it to a range of weaker acids (predominantly acetic and propionic acids), although this hypothesis is not always supported by clinical evidence.

GRAM-NEGATIVE RODS

The majority of aerobic or facultatively anaerobic Gram-negative rods fall into the genus *Haemophilus*. These organisms were not detected in early studies until, as with the enterococci, appropriate isolation media were used. These provided all of the essential growth factors required by members of this genus including haemin (X-factor) and nicotinamide adenine dinucleotide (V-factor). Haemophili were then found to be commonly present in saliva, on epithelial surfaces, and in dental plaque. The five species that were regularly isolated together with their growth factor requirements, are: *H. influenzae* (X- and V-factor), *H. parainfluenzae* biotypes I, II and III (V-factor only), *H. segnis* (V), *H. paraphrophilus* (V), and *H. aphrophilus* (requiring neither X nor V). The distribution of individual species varies with the oral surface sampled but as yet there is no information about the adhesive mechanisms involved. Strains have been isolated from jaw infections and cases of infective endocarditis but, in general, the pathogenic potential of these organisms is low.

Other facultatively anaerobic Gram-negative rods have been identified as *Eikenella corrodens*. Colonies of this species characteristically pit the surface of agar plates. Strains of *E. corrodens* have been isolated from a range of oral infections including endocarditis and abscesses, and have been implicated in periodontal disease. *Capnocytophaga* is a relatively new genus of CO_2-dependent Gram-negative rods, which have a gliding motility. They were originally classified as *Bacteroides ochraceus*. These organisms are found in sub-gingival plaque, and increase in proportions in gingivitis. Three species have been recognized (*C. gingivalis*, *C. ochracea*, *C. sputigena*), but few simple tests are available at present to differentiate them. *Capnocytophaga* are opportunistic pathogens and have been isolated from a number of infections in immunocompromised patients. *Actinobacillus actinomycetemcomitans* is another capnophilic (CO_2-liking) species that is found more commonly in certain forms of peridontal disease than in plaque from healthy sites. This species produces a powerful leucotoxin and is closely associated with juvenile peridontitis, a particularly aggressive form of peridontal disease that can affect young adults (see Chapter 7). The gliding bacterium *Simonsiella* has been isolated from epithelial surfaces of the oral cavity of man and a variety of animals. These organisms have a unique cellular morphology being composed of unusually large, multi-cellular filaments in groups, or multiples of eight cells. At one time these organisms were implicated with lichen planus but a recent study was unable to confirm such an association.

Obligately anaerobic Gram-negative rods comprise a large proportion of the microflora of dental plaque. However, the taxonomy and classification of many of these organisms is confused due to several inter-related factors. Many isolates are pleomorphic and problems can arise in making the fundamental decision as to whether an organism is a Gram-negative coccus or rod. Tests that were successful for aerobic or facultatively anaerobic Gram-negative rods were applied initially to the obligate anaerobes, but in many cases these were shown subsequently to be inappropriate. For example, some strains grow so poorly that clear results in fermentation tests are not obtained, and the concentration of metabolites is too low to be analysed satisfactorily. As a consequence, difficulties can arise in discriminating between a strain that is growing poorly in a range of tests and one that is truly asaccharolytic. The development and application of new tests such as lipid analyses and enzyme mobilities have been necessary for the speciation of isolates that appeared non-reactive by conventional methods. Confusion over the taxonomy of this group of bacteria has been increased through the realization that some reference strains, on whose properties divisions were originally based, were atypical in some cases.

The majority of oral anaerobic Gram-negative rods were placed in the genus *Bacteroides*. However, recent chemotaxonomic studies have shown that this genus should have a much narrower definition which will restrict its members to those of the *B. fragilis*-group found predominantly in the gut. Some of the commonly isolated oral organisms produce colonies with a characteristic brown or black pigment when grown on blood agar. These pigments, originally thought to be melanin, have been identified as protoporphyrin and protohaem, and are required for the synthesis of cytochrome b and for the respiratory quinones (menaquinones), respectively. These organisms were referred to collectively as black-pigmented *Bacteroides* (Table 3.8), but a more appropriate term would now be black-pigmented anaerobes. Asaccharolytic organisms have been placed in the new genus, *Porphyromonas*, while those that are saccharolytic are in another new genus, *Prevotella* (Table 3.8).

Table 3.8 Past and present nomenclature of human oral obligate anaerobes formerly classified as *Bacteroides* species

Brown/black pigmenting		Non-pigmenting	
Present nomenclature	*Previous nomenclature*	*Present nomenclature*	*Previous nomenclature*
Porphyromonas gingivalis	*Bacteroides gingivalis*	*Prevotella buccae*	*Bacteroides buccae*
Porphyromonas endodontalis	*Bacteroides endodontalis*	*Prevotella buccalis*	*Bacteroides buccalis*
		Prevotella heparinolytica	*Bacteroides heparinolyticus*
Prevotella melaninogenica	*Bacteroides melaninogenicus*	*Prevotella oralis*	*Bacteroides oralis*
Prevotella intermedia	*Bacteroides intermedius*	*Prevotella oris*	*Bacteroides oris*
Prevotella loescheii	*Bacteroides loescheii*	*Prevotella oulora*	*Bacteroides oulorum*
Prevotella denticola	*Bacteroides denticola*	*Prevotella veroralis*	*Bacteroides veroralis*
		Prevotella zoogleoformans	*Bacteroides zoogleoformans*
		?*	*Bacteroides capillosus*
		?	*Bacteroides forsythus*
		?	*Bacteroides gracilus*
		?	*Bacteroides pneumosintes*
		?	*Bacteroides ureolyticus*

?* The previous nomenclature remains at present, but these organisms will not be retained within the genus *Bacteroides* in the future.

Following these recent major changes in taxonomy, the primary habitats of members of these new genera has now been elucidated. *Porphyromonas asaccharolytica* has been reported only occasionally from the mouth and is usually found in a wide range of non-oral sites, especially the gut. In contrast, *Porphyromonas gingivalis* is found almost solely at sub-gingival sites (especially in deep periodontal pockets) although it has also been recovered from the tongue and tonsils. It is highly virulent in experimental infection studies in animals and produces a range of putative virulence factors including proteases, cytotoxic metabolic products, and a capsule (Chapter 7). *Porphyromonas endodontalis* has an even more restricted habitat being recovered mainly from infected root canals. Both *P. gingivalis* and *P. endodontalis* are rarely found in health and their source of origin is a cause of much debate. Some have proposed that they may be derived exogenously from other subjects, while others believe that they may be present in reservoirs on mucosal surfaces such as the tongue or tonsils. It may be that they are present in dental plaque in health but in such low numbers as to be below the sensitivity of routine detection methods. When the environment changes in disease, then their proportions (and hence their apparent prevalence) would increase. Phenotypically related species have been isolated from a number of animals including '*Bacteroides levii*' (cattle), '*B. salivosus*' (cats) and '*B. macacae*' (monkeys), while *P. gingivalis*-like organisms have been isolated from lesions of broken mouth in sheep. The precise taxonomic position of these species has still to be determined.

The new definition of the genus *Bacteroides* has meant that members of the '*B. melaninogenicus-B. oralis*' group have been placed in a new genus, *Prevotella* (Table 3.8). Species within this group are moderately saccharolytic, producing acetic and succinic acids from glucose. This new genus includes the pigmented species *P. intermedia* (formerly *B. intermedius*), *P. melaninogenica* (formerly *B. melaninogenicus*), *P. loescheii*, *P. corporis* (this species is closely related to *P. intermedia*, but is generally isolated from non-oral sites) and some strains of *P. denticola*. The oral non-pigmented species include *P. buccae*, *P. buccalis*, *P. heparinolytica*, *P. oralis*, *P. oris*, *P. oulora*, *P. veroralis* and *P. zoogleoformans*. The majority of these species can be isolated on occasions from dental plaque, particularly from sub-gingival sites. Some species are associated with disease and increase in numbers and proportions during periodontal disease, and have also been recovered from abscesses (Chapters 7 and 8).

There are other saccharolytic, non-pigmented species that no longer fall within the new definition of the genus *Bacteroides*. The taxonomic fate of these organisms is uncertain at present, but they will remain with their original epithet until their classification is resolved. These species include '*B. capillosus*', '*B. forsythus*', '*B. gracilis*' and '*B. pneumosintes*' (Table 3.8). Some species have unusual properties; for example, the growth of '*B. gracilis*' is stimulated by formate or hydrogen and fumarate or nitrate. *P. zoogleoformans* grows as a viscous, ropy mat in liquid culture while '*B. ureolyticus*' and '*B. forsythus*' can pit the surface of blood agar plates. 'B. gracilis' and 'B. ureolyticus' do not possess flagella but move due to an unusual twitching motility. Members of these bacterial groups can be distinguished by patterns of metabolic end products, fermentation profiles, and the production of certain constitutive enzymes. Some species are also found in the oral cavity of animals, e.g. 'B. rectum' from dogs and cats.

Another major group of obligately anaerobic Gram-negative bacteria belong to the genus *Fusobacterium*. Strains are asaccharolytic and produce predominantly butyric acid, although they do incorporate glucose into cell constituents. Cells are characteristically in the form of long filaments or pleomorphic rods and include the following oral species: *F. alocis* and *F. sulci* from the normal gingival crevice and *F. periodonticum* from sites with periodontal disease. The most commonly isolated species, however, is *F. nucleatum*. Recently, several subspecies have been recognized within *F. nucleatum*: subspecies *nucleatum*, subsp. *polymorphum* and subsp. *vincentii* (also referred to by some as subsp. *fusiforme*). Interestingly it has been proposed that these subspecies may have different associations with health and disease. *F. nucleatum* subsp. *polymorphum* was most commonly isolated from the normal gingival crevice whereas subspecies *nucleatum* was recovered mainly from periodontal pockets. *F. simiae* has been isolated from the mouth of macaque monkeys.

Other oral Gram-negative anaerobic and micro-aerophilic bacteria include *Leptotrichia buccalis* which are filamentous cells with a pointed end. They can be differentiated from *Fusobacterium* spp. by the production of lactate as the major fermentation product of *Leptotrichia* spp. *Mitsuokella dentalis* has been isolated from infected root canals; some strains possess a thick capsule which may contribute to their pathogenicity. *Wolinella succinogenes* is asaccharolytic and is a formate/fumarate-requiring species; *Campylobacter concisus* and *C. sputorum* are also isolated from the mouth occasionally. *Wolinella curva* and *W. recta* have now been reclassified as *Campylobacter curvus* and *C. rectus*, respectively. These strains are not always easily identified by conventional tests and other methods are under investigation, such as whole cell protein profiles. Some of these species have flagella and are motile. The *Wolinella* and *Campylobacter* species have a single flagellum, while *Selenomonas sputigena* is a curved to helical bacillus with a tuft of flagella. *Selenomonas noxia*, *S. flueggei*, *S. infelix*, *S. dianae* and *S. artemidis* are new species that have been recently described in plaque from the human gingival crevice. Another helical or curved Gram-negative oral anaerobe is *Centipeda periodontii* which has numerous flagella which spiral around the cell. There has been one report of *Helicobacter* (formerly *Campylobac-*

ter) *pylori* from dental plaque; this species is usually isolated from the stomach where it is associated with gastritis.

Spirochaetes are numerous in sub-gingival plaque and can readily be detected using dark-field or electron microscopy. Several morphological types can be distinguished according to cell size and the arrangement of endoflagella. The numbers of spirochaetes are raised in periodontal disease, but whether they cause disease or merely increase following infection is not clear. As yet, no clear correlation has been found between any particular morphotype and the severity of disease. Oral spirochaetes fall within the genus *Treponema* and a number of species have been proposed, including *T. denticola, T. macrodentium, T. oralis, T. scoliodontium, T. socranskii* and *T. vincentii. T. pectinovorum* is found in non-human primates. To date, little is known about the physiology of these organisms because of difficulties associated with their laboratory cultivation. Gradually, such problems are being resolved and progress is now being made in this area. Some of the medium- and small-sized spirochaetes have been grown successfully. *T. denticola* has been shown to secrete extracellular and degradative enzymes including proline aminopeptidase and a trypsin-like enzyme; the latter can also degrade collagen. *T. denticola* is asaccharolytic while *T. socranskii* can ferment carbohydrates to acetic, lactic and succinic acids.

MYCOPLASMA

Mycoplasma are pleomorphic bacteria which possess an outer membrane that is not rigid. They can be grown either on enriched nutrient media or in tissue culture. Originally, the mycoplasma-group included 'L' forms, but many of the latter revert to bacteria and thus are not part of the group. *Mycoplasma* have been isolated from saliva, the oral mucosa, and dental plaque, but their role at these sites is not clear. A considerable amount of work remains to be done on the taxonomy of the oral mycoplasma group. The poorly defined *Mycoplasma salivarium* and the more established *M. pneumoniae* and *M. hominis* have been isolated from saliva and the oral mucosa; other oral species include *M. orale* and *M. buccale*. Estimates of the carriage rate of *Mycoplasma* from the oral cavity vary from 6% to 32% of the population in the limited surveys completed.

FUNGI

Fungi form a small but not insignificant part of the oral microflora. The 'perfect fungi' (fungi that divide by sexual reproduction) are thought to be only transients in the oral cavity, although *Aspergillus* and *Mucor* spp. have been reported in the oral microflora in limited studies of populations in the Indian subcontinent. The 'imperfect fungi' (which divide by asexual reproduction) are usually yeasts, and form part of the resident oral microflora.

The largest proportion of the fungal microflora is made up of *Candida* spp. *Candida albicans* is by far the most common isolate, but another 27 *Candida* species have been isolated from the mouth. The next most commonly isolated species are *Candida glabrata* (formerly 'Torulopsis glabrata'), *Candida tropicalis*, *Candida krusei*, *Candida parapsilosis*, *Candida guilliermondi* and, on occasions, *Rhodotorula* and *Saccharomyces* spp.

Estimations of carriage rates of *Candida* species in the mouth are difficult to interpret because of the variation in isolation techniques. The use of swabs has been shown to be inaccurate. Techniques such as foam imprint cultures, which use a piece of foam to sample a mucosal site, and gargle techniques are much more accurate. Data from a variety of surveys have shown that carriage rates may vary from 2% to 71% of asymptomatic adults. Carriage rates increase, and can approach 100%, in medically compromised patients or those on broad spectrum antibacterial agents.

Candida are distributed evenly throughout the mouth. The most common site of isolation is the dorsum of the tongue, particularly the posterior area near the circumvallate papillae. The isolation of *Candida* is influenced by the presence of intra-oral devices such as plastic dentures or orthodontic appliances, particularly in the upper jaw on the fitting surface. Plaque has also been shown to contain *Candida* spp., but the exact proportion and significance of the yeasts in health and disease is unclear. There is agreement that the mouth is the body site where *Candida* spp. are most prevalent. The mouth is thought to be the source of yeast colonization of the gut. Saliva is the vehicle for the transmission of *Candida* spp. to other areas of the body.

Colonization of the mouth by yeasts is thought to occur either at birth or soon afterwards. The carriage rate falls in early childhood and increases during middle and later life. Diurnal variation of candidal counts has been described, but this may be a reflection of the inaccuracy of isolation techniques rather than a consistent finding.

VIRUSES

The detection of viruses in the oral cavity has until recently been difficult. This difficulty was caused by the limitation of the two main detection methods: electron microscopy and growth in tissue culture. Growth in tissue culture is laborious and time-consuming and, once grown, the virus still has to be identified. Electron microscopy is also time-consuming, and allows only presumptive identification based on morphology. The major advances in virology have been two-fold: the development of specific antisera to viruses and the polymerase chain reaction (PCR). A wide variety of specific antisera to oral viruses are now available. These allow direct identification of viral antigens using radioactive or visual probes. They also allow indirect identification through the detection of general viral antibody in saliva, mucosal tissue, or serum.

The use of specific antisera has been important to virologists but, to a large extent, this has been overshadowed by PCR techniques. This process amplifies copies of the viral genome present in tissue, saliva or serum. The amplified viral genome can be identified by matching it (homology) to the known genome using hybridization techniques. These are highly sensitive and, in theory, can be used to identify a single genome present in the original sample.

The most common virus detected in the oral cavity is *Herpes simplex*. Both type 1 and type 2 have been isolated from the mouth, but type 1 is the most common. *Herpes simplex* type 1 is the cause of cold sores. The virus can be detected occasionally in the oral cavity in the absence of cold sores and is, therefore, probably persistent. The virus can also remain latent. It migrates rapidly along the trigeminal nerve to the ganglion where it remains latent until reactivated. The precise cause of reactivation is still not clear, but UV light and stress appear to be involved. Once reactivated, the genome passes back down the peripheral nerve to cause the characteristic cold sores. These cold sores rupture to release further *Herpes simplex* type 1 virus particles.

Cytomegalovirus is present in most individuals. It has been detected in the saliva of symptomless adults, but its portal of entry into the oral cavity is not clear. *Coxsackie virus* A2, 4, 5, 6, 8, 9, 10, and 16 have all been detected in saliva and in oral epithelium. The detection of these viruses has usually been secondary to, or in the primary phase of, an infection of hand, foot and mouth disease or herpangina. A variety of *Papilloma* viruses has been isolated from the oral cavity, usually in association with small warts. These warts have been implicated with oral cancer, and are common in patients with AIDS.

Hepatitis and *Human Immunodeficiency Virus* (HIV) can be found in the oral cavity, especially in saliva, where their presence poses a significant cross-infection threat. Both groups of viruses are discussed in detail in Chapter 9. Other viruses that have been detected in the oral cavity are measles and mumps, but usually in association with oral lesions.

PROTOZOA

There is little definitive information about the presence of protozoa in the oral microflora. *Entamoeba gingivalis*, *Trichomonas tenax* and *Hamblia* spp. have been isolated from the oral cavity of some asymptomatic patients, but their exact prevalence and function in the oral cavity is still unclear.

SUMMARY

The mouth supports the growth of a wide diversity of micro-organisms including bacteria, yeasts, viruses, and (on occasions) protozoa. Many of the bacteria are fastidious in their nutritional requirements and are difficult to grow and identify in the laboratory; many are also obligate anaerobes.

Table 3.9 Bacterial genera found in the oral cavity

	Gram-positive	*Gram-negative*
Cocci	*Enterococcus*	*Branhamella*
	Peptostreptococcus	*Neisseria*
	Streptococcus	*Veillonella*
	Stomatococcus	
Rods	*Actinomyces*	*Actinobacillus*
	Bifidobacterium	*(Bacteroides)**
	Corynebacterium	*Campylobacter*
	Eubacterium	*Capnocytophaga*
	Lactobacillus	*Centipeda*
	Propionibacterium	*Eikenella*
	Rothia	*Fusobacterium*
		Haemophilus
		Leptotrichia
		Mitsuokella
		Porphyromonas
		Prevotella
		Selenomonas
		Simonsiella
		Treponema
		Wolinella

* The genus *Bacteroides* has been redefined. In time, all oral bacteria previously placed in this genus will be reclassified.

Numerous problems associated with the classification of oral bacteria have been resolved in recent years due to the application of chemotaxonomic techniques. New species of streptococci and *Actinomyces* have been recognized and shown to have a specific habitat and niche. Some species of streptococci that were previously grouped together have now been shown to differ markedly in their ability to colonize the tooth surface and in their distribution on the oral mucosa. Likewise, rapid advances have been made in the taxonomy of obligately anaerobic Gram-negative rods, many of which are asaccharolytic and were difficult to distinguish using conventional tests. New criteria include lipid analyses, enzyme mobilities, and the breakdown of natural substrates. The benefits of these changes in classification are that closer associations of individual species with sites in health and disease are being discerned.

Bacteria are the predominant components of the resident oral microflora; a list of genera is given in Table 3.9. The high species diversity reflects the wide range of nutrients available endogenously, the varied types of habitat for colonization, and the opportunity provided by biofilms such as plaque for survival on surfaces. Despite this diversity, it is interesting to note that many micro-organisms commonly isolated from neighbouring ecosystems, such as the skin and the gut, are not found in the mouth. This emphasizes the unique and selective properties of the mouth for microbial colonization.

FURTHER READING

Austin, B. and Priest, F. G. (1986) *Modern Bacterial Taxonomy*. Van Nostrand Reinhold, Wokingham.

*Balows, A., Truper, H. G., Dworkin, M., Harder, W. and Schleifer, K. H. (eds.) (1991) *The Prokaryotes. A handbook on the biology of bacteria*. 2nd edn. Springer-Verlag, New York.

Beighton, D., Hardie, J. M. and Whiley, R. A. (1991) A scheme for the identification of viridans streptococci, *Journal of Medical Microbiology*, **35**, 367–72.

Delwiche, E. A., Pestka, J. J. and Tortorello, M. L. (1985) The Veillonellae: Gram-negative cocci with a unique physiology. *Annual Reviews of Microbiology*, **39**, 175–93.

Hamada, S., Michalek, S. M., Kiyono, H., Menaker, L. and McGhee, J. R. (eds.) (1986) *Molecular Microbiology and Immunobiology of Streptococcus mutans*. Elsevier, Amsterdam.

Hardie, J. M. (1989) Application of chemotaxonomic techniques to the taxonomy of anaerobic bacteria. *Scandinavian Journal of Infectious Disease* Suppl., **62**, 7–14.

Kilian, M. and Schiott, C. R. (1975) Haemophili and related bacteria in the human oral cavity. *Archives of Oral Biology*, **20**, 791–6.

Mayrand, D. and Holt, S. C. (1988) Biology of asaccharolytic black-pigmented *Bacteroides* species. *Microbiological Reviews*, **52**, 134–52.

*Parker, M. T. and Duerden, B. I. (eds.) (1990) *Topley & Wilson's Principles of Bacteriology, Virology and Immunity. Volume 2. Systematic Bacteriology*. Edward Arnold, London.

Samaranayake, L. P. and MacFarlane, T. W. (1990) *Oral Candidosis*. Wright, London.

* These volumes contain detailed chapters on the classification, identification, and properties of most of the major groups of bacteria found in the oral cavity.

4 Acquisition, adherence, distribution and functions of the oral microflora

The foetus in the womb is normally sterile. During delivery the baby comes into contact with the normal microflora of the mother's uterus and vagina, and at birth with the micro-organisms of the atmosphere and of the people in attendance. Despite the widespread possibility of contamination, the mouth of the newborn baby is usually sterile. However, from the first feeding onwards, the mouth is regularly inoculated with micro-organisms and the process of the acquisition of the resident oral microflora begins.

ACQUISITION OF THE RESIDENT ORAL MICROFLORA

Acquisition depends on the successive transmission of micro-organisms to the site of potential colonization. Initially, in the mouth, this is by passive contamination from the mother, from food, milk and water, and from the saliva of individuals in close proximity to the baby. Acquisition of micro-organisms from the birth canal itself may be of only limited significance. Studies of the transmission of candida and lactobacilli from the vagina of the mother to the newborn infant suggest that these organisms do not usually become established initially as part of the resident oral microflora of the newborn, although they may be present transiently. In contrast, the role of saliva in the process of acquisition has been confirmed conclusively. Bacteriocin-typing of strains has enabled the transfer of *S. salivarius* and mutans streptococci from mother to child via saliva to be followed. Similarly, comparisons of the genetic material (DNA) of a variety of oral bacterial species (using restriction endonuclease mapping techniques) have shown that the same digest pattern (and hence presumably the same strain) is found within family groups and that different patterns are observed between such groups.

Pioneer community and ecological succession

The mouth is highly selective for micro-organisms even during the first few days of life. Only a few of the species common to the oral cavity of adults, and even less of the large number of bacteria found in the environment, are able to colonize the mouth of the newborn. The first micro-organisms to colonize are termed **pioneer species**, and collectively they make up the **pioneer microbial community**. These pioneer species continue to grow and colonize until environmental resistance is encountered. This can be due to several limiting forces which act as barriers to further development. These barriers include both physical and chemical factors. In the oral cavity physical factors including the shedding of epithelial cells (desquamation) and the shear forces from chewing and saliva flow. Nutrient requirements, Eh, pH, and the antibacterial properties of saliva can act as chemical barriers limiting growth.

During the development of the pioneer community one genus or species is usually predominant. In the mouth, the predominant organisms are streptococci and in particular, *S. salivarius*. With time, the metabolic activity of the pioneer community modifies the environment. This provides conditions suitable for colonization by a succession of other populations, for example, by changing the Eh or pH, by modifying or exposing new receptors for attachment ('cryptitopes', Chapter 5), or by generating novel nutrients, for example, as end products of metabolism (lactate, succinate, etc.) or as breakdown products (peptides, haemin, etc). In this way the pioneer community can influence the pattern of microbial succession. This involves the progressive development of a pioneer community (containing few species) through several stages in which the number of microbial groups increase. Eventually a stable situation is reached with a high species diversity; this is termed the **climax community** (Figure 4.1). Succession is associated with a

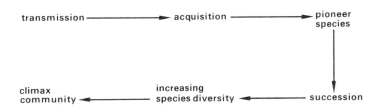

Figure 4.1 Ecological stages in the establishment of a microbial community.

change from a site possessing few niches to one with a multitude of potential niches. A climax community reflects a highly dynamic situation and must not be regarded as a static state.

The oral cavity of the newborn contains only epithelial surfaces for colonization. The pioneer populations consist of mainly aerobic and facultatively anaerobic species. *Streptococcus salivarius* can be isolated from the mouth of infants as early as 18 hours after birth. Indeed, streptococci are the dominant pioneer species in the oral cavity over the first year of life. In contrast, lactobacilli may be present only transiently during this period.

The diversity of the pioneer oral community increases during the first few months of life. The isolation frequency of some oral organisms from the mouth of 44 twelve-month-old infants is shown below:

Isolated from all infants	*Streptococcus, Staphylococcus, Neisseria, Veillonella*
Isolated from >50% infants	*Actinomyces, Lactobacillus, Rothia, Fusobacterium*
Isolated from <50% infants	*'Bacteroides'* (probably *Prevotella*), *Leptotrichia, Candida, Corynebacterium*

The isolation frequencies of *Actinomyces*, *Fusobacterium* and other anaerobic species increase following tooth eruption.

Allogenic and autogenic succession

The development of a climax community at an oral site can involve examples of both allogenic and autogenic succession. In allogenic succession, factors of non-microbial origin are responsible for an altered pattern of community development. Species such as mutans streptococci and *S. sanguis* only appear in the mouth once teeth have erupted. In the cases in which they have been isolated from the edentulous (toothless) mouth, these species are associated with the insertion of artificial devices such as acrylic obturators in children with cleft palates. Similar differences occur in the distribution of *Actinomyces* in the mouth of young children. Strains resembling *A. naeslundii* were isolated from 40% of pre-dentate infants, but *A. viscosus* did not appear until after tooth eruption and its frequency of isolation increased with age. These observations suggest these species prefer to colonize hard surfaces.

The increase in number of obligate anaerobes once teeth are present is an example of autogenic succession in which community development is influenced by microbial factors. The metabolism of the aerobic and facultatively anaerobic pioneer species can lower the redox potential in plaque and create conditions suitable for colonization by strict anaerobes (Chapter 5). Other examples of autogenic succession would be the development of food chains (also termed foodwebs) whereby the metabolic end product of one organism becomes a primary nutrient source for a second. A further example is the exposure of new receptors on host macromolecules for bacterial adhesion ('cryptitopes'; Chapter 5).

AGEING AND THE ORAL MICROFLORA

The acquisition of the oral microflora continues with age. Following tooth eruption, the isolation frequency of spirochaetes and black-pigmented anaerobes increases. For example, the latter group of organisms are recovered from 18–40% of children aged five years but are found in over 90% of teenagers aged 13–16 years. It has been proposed that the increased prevalence of both bacterial groups during puberty might be due to hormones entering the gingival crevice and acting as a nutrient source. The rise in *Prevotella intermedia* in plaque during the second trimester of pregnancy has also been ascribed to the elevated serum levels of oestradiol and progesterone which can satisfy the naphthaquinone requirement of this organism. Rises in black-pigmented anaerobes have also been observed in women taking oral contraceptives. However, in a recent study using dark-field microscopy, spirochaetes were detected in the sub-gingival microflora of 66% of 77 pre-pubertal Dutch children. This implies that hormonal changes cannot be the only factor affecting the prevalence of these fastidious bacteria.

In adults, the resident oral microflora remains relatively stable and coexists in reasonable harmony with the host. This stability (termed microbial homeostasis, Chapter 5) is not a passive response to the environment, but is due to a dynamic balance being achieved from numerous inter-bacterial and host–bacterial interactions.

Some variations in the oral microflora have been discerned in later life and can be attributed to both direct and indirect effects of ageing. In the case of the latter, variations can occur if the habitat or environment is severely perturbed. For example, the risk of cancer rises with age, and cytotoxic therapy or myelosuppression combined with the disease itself is associated with the increased carriage of *C. albicans* and non-oral opportunistic pathogens such as enterobacteria (*Klebsiella* spp., *Escherichia coli*, *Pseudomonas aeruginosa*) and *Staphylococcus aureus*. The incidence of the wearing of dentures also increases with age and this also promotes colonization by *C. albicans*. Information is now beginning to emerge on direct age-related changes in the oral microflora. Significantly higher proportions and isolation frequencies of lactobacilli and staphylococci (mainly *S. aureus*) in saliva were found in healthy subjects aged 70 years or over while yeasts were isolated more often and in higher numbers from saliva in those aged 80 years and over. In plaque, *A. viscosus* was found in higher proportions than *A. naeslundii* in those over 60 years of age. It has been proposed that cell-mediated immunity declines with age, but the precise effect of old ageing on the innate and specific host defences of the mouth has yet to be established definitively. The incidence of oral candidosis is more common in the elderly and this has been attributed to not only the increased likelihood of denture wearing but also to physiological changes in the oral mucosa, malnutrition, and to trace element deficiencies. There have been reports of increased isolations of enterobacteria from the oro-pharynx of the elderly, but this seems to be related in many cases to the health of the individual rather than to their age *per se*, with the highest incidences being in the most debilitated individuals. One of the fundamental problems in determining whether the oral microflora changes in old age is that the chronological age of a person does not always equal their physiological age!

Social habits can perturb the balance of the oral microflora. The regular intake of dietary carbohydrates can lead to the enrichment of aciduric (acid-tolerant) and cariogenic species such as mutans streptococci and lactobacilli. Smoking has been shown to reduce bacterial counts, especially those of *Neisseria* spp., on a variety of oral surfaces, and is also a risk factor for periodontal diseases. Also, mutans streptococci *S. sanguis* are not detected in the mouths of individuals who have full dentures when these surfaces are not worn, although both groups of bacteria can reappear when these 'hard surfaces' are inserted again.

Summary

The acquisition of the normal oral microflora follows a specific ecological progression from a small number of pioneer species (the predominant species is usually *S. salivarius*) to a complex climax community. This development involves both allogenic and autogenic succession. In allogenic succession, community development is influenced by non-microbial factors; microbial factors are responsible for autogenic succession.

METHODS OF DETERMINING THE RESIDENT ORAL MICROFLORA

Problems are encountered in determining the composition of the microbial communities at various sites in the mouth. These relate to the basic difficulties of removing the majority of the micro-organisms (many of which are of necessity bound tenaciously to a surface or to each other), their transport to the laboratory in a viable state, and then their enumeration and identification (Table 4.1). The main stages in sample processing are illustrated in Figure 4.2.

Table 4.1. Some properties of the oral microflora that contribute to the difficulty in determining its composition

Property	Comment
High species diversity	The oral microflora, and especially dental plaque, consists of a diverse number of microbial species, some of which are present only in low numbers
Surface attachment/co-aggregation	Oral micro-organisms attach firmly to surfaces and to each other, and therefore have to be dispersed without loss of viability
Obligate anaerobes	Many oral bacteria lose their viability if exposed to air
Fastidious nutrition	Some bacteria are difficult to grow in pure culture and may require specific cofactors etc. for growth. Some groups (e.g. spirochaetes) cannot be cultured
Slow growth	The slow growth of some organisms (e.g. 14 days) makes enumeration time-consuming
Identification	The classification of many oral micro-organisms remains unresolved or confused; simple criteria for identification are not always available (particularly for many obligate anaerobes)

Sample taking

The microflora can vary in composition over relatively small distances. Therefore, large plaque samples or a number of smaller samples from different sites but which are pooled together are of little value because important site differences will be obscured. Consequently, small sam-

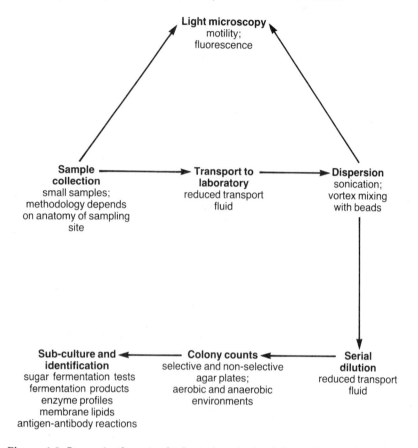

Figure 4.2 Stages in the microbiological analysis of the oral microflora.

ples from discrete sites are preferable, but the method of sampling will depend on site to be studied.

The oral mucosa can be sampled by swabbing, direct impression techniques, or by removing epithelial cells by scraping or scrubbing with a blunt instrument into a container. Data can then be related to a fixed area or to an individual epithelial cell. Saliva can be collected by expectoration into a sterile container; the saliva flow can be at a normal resting rate (i.e. unstimulated) or it can be stimulated by chemical means or by chewing. Although a greater volume is collected by stimulation, such samples will also contain many more organisms that have been dislodged from oral surfaces.

There is no universally accepted way of sampling dental plaque. The accessible smooth surfaces of enamel pose few problems and a range of dental instruments have been used. It is more difficult to remove plaque from approximal surfaces between teeth although dental probes, scalers, dental floss and abrasive strips have all been used with varying degrees of success. Pits and fissures are also difficult to sample and the amount of plaque removed can be dependent on the anatomy of the site. Fine probes, pieces of wire, blunt hypodermic needles, and tooth-picks have all been used successfully. Sub-gingival plaque has proved to be the most difficult to sample because of the inaccessibility and anaerobic nature of the site. High numbers of obligately anaerobic bac-teria are found in the gingival crevice and periodontal pocket, most of which will lose their viability rapidly if exposed to air. In disease, the anatomy of the site means that those organisms at the base of the pocket, near the advancing front of the lesion, are likely to be of most interest (Chapter 7). Again, it is important not to remove plaque from other sites in the pocket, as this might obscure significant relationships between specific bacteria and disease. To overcome these problems a number of methods have been developed, all of which have their par-ticular advantages and drawbacks. For example, a simple approach has been to insert paper points into pockets but the number of firmly adherent organisms removed from the root of the tooth will be small. Samples have also been taken by irrigation of the site and retrieval of the material through syringe needles; however this method will obviously remove plaque from the whole depth of the pocket. A particularly sophisticated method employs a broach kept withdrawn in a cannula which is flushed constantly with oxygen-free nitrogen. The broach is used to sample plaque only when the cannula is in position near the base of the pocket. After sampling, the broach is retracted into the cannula and withdrawn. Perhaps the most frequent approach has been to use a curette or scaler after the supra-gingival area has been cleared. The scaler tips can be detached and placed immediately in gas-flushed tubes containing reduced (anaerobic) transport fluid for delivery to the laboratory. Alternatively, when periodontal surgery is needed, plaque has been removed from extracted teeth or from surfaces exposed when 'gingival flaps' are reflected. It is impossible to design experiments to compare the efficiency of each of these sampling methods, but it is important to realize, particularly when comparing studies in which different sampling procedures have been used, that the results will, to a certain extent, reflect the method adopted.

Transport and dispersion

All samples need to be transported to the laboratory for processing as quickly as possible. Specially designed transport fluids help to reduce the loss of viability of some of the more delicate organisms during delivery to the laboratory. These fluids usually contain reducing agents to maintain a low redox potential, thus helping to preserve the obligate anaerobes.

Clumps and aggregates of bacteria must be dispersed efficiently (ideally to single cells) if the specimen is to be diluted and counted accurately. Plaque poses a particular problem in this respect because, by definition, it is a complex mixture of a range of micro-organisms bound tenaciously to one another. It is now accepted that mere vortex mixing of a sample is inadequate. Mild sonication produces the maximum number of particles from a specimen but it exerts a selective effect by specifically damaging spirochaetes and some other Gram-negative bacteria, particularly *Fusobacterium* species. One of the most efficient methods, particularly for sub-gingival plaque, is to vortex samples with small glass beads in a tube filled with carbon dioxide.

Cultivation

Once dispersed, samples are usually serially diluted in transport fluid and aliquots are spread on to a number of freshly prepared, pre-reduced agar plates. These plates are chosen to grow either the maximum number of bacteria (generally, various forms of blood agar are used for this purpose) or, in order to encourage the growth of some of the minor components of the microflora, a number of selective media have been devised which permit the growth of only a limited number of species. For example, the addition of vancomycin to blood agar plates will inhibit most Gram-positive bacteria, while a high sucrose concentration encourages the growth of oral streptococci, and plates with a low pH favour lactobacilli. It should be stated that these media are selective and not specific. The identity of the colonies on these plates must be confirmed; their colonial appearance or growth on a particular medium should not be regarded as diagnostic. Also, some bacteria will not grow unless additional factors are added to the medium. Haemophili were only recovered when the appropriate co-factors necessary for their growth were added to the isolation media. Depending on the bacteria being cultivated, plates have to be incubated for different times and under different atmospheric conditions. For example, to grow some obligate anaerobes, plates will need 7–14 days incubation at 37°C in an anaerobic jar or cabinet filled with a gas mix containing $CO_2/H_2/N_2$; in contrast, *Neisseria* require only two days incubation in air. Some organisms grow optimally in 10% CO_2 in air.

Enumeration and identification

Colonies are counted and their concentration in the original sample determined by compensating for the dilution steps. Representative colonies can then be sub-cultured to check for purity and for subsequent identification. This process assumes that:

(a) Cells of the same micro-organism produce colonies with an identical morphology.
(b) Cells of different species produce distinct morphologies.
(c) One colony arises from a single cell.

Generally, these assumptions hold true except for (c), as most colonies inevitably arise from small aggregates of cells; this emphasizes the need for efficient dispersion of samples. The first level of discrimination involves the Gram-staining of subcultured colonies; bacteria are then grouped according to whether their cells are Gram-positive or Gram-negative, and are rod- or coccal-shaped. This dictates which tests will be necessary to achieve speciation. Some bacteria can be identified using simple criteria – for example, sugar fermentation tests – while others require a more sophisticated approach such as the application of gas-liquid chromatography to determine their acid end-products of metabolism (Table 3.1). The use of probes (antibodies or DNA) on colonies is now being used to speed up identification. Many other issues relating to bacterial identification were discussed in the previous chapter.

Microscopy

As an alternative to many of the lengthy steps associated with conventional culture techniques (Figure 4.2) an idea of the principal morphological groups of bacteria can be obtained using light microscopy. Dark-field illumination or phase-contrast techniques have been used to quantify directly the numbers of motile bacteria in dental plaque, including spirochaetes, which are extremely difficult to cultivate by conventional means. As the numbers of spirochaetes and motile organisms have been related to the severity of some periodontal diseases, and because microscopy is relatively cheap and gives results quickly, it has been hoped that such techniques could be used in the clinic to monitor the progress of patients undergoing treatment. However, a major disadvantage of dark-field and phase contrast microscopy is that most of the putative pathogens cannot be recognized by morphology alone. To overcome this problem, antisera (monoclonal or specific polyclonal) have been raised against a limited number of bacteria that are implicated in, or act as markers of, disease. These antisera can be either conjugated with a fluorescent dye (direct fluorescence) or used in conjunction with commercially prepared, conjugated anti-rabbit IgG (indirect fluorescence) to quantify rapidly the approximate numbers of selected key bacteria in samples, particularly dental plaque. Gene (DNA) probes are also being used for similar purposes.

Scanning and transmission electron microscopy has proved useful in studying plaque formation, and it has also been used to show that bacteria invade gingival tissues in aggressive forms of periodontal disease. Electron-dense markers (ferritin, peroxidase, gold granules) conjugated with antibodies can label specific surface antigens on bacteria and facilitate their identification or location on a surface or within plaque.

In situ models

As a result of some of the methodological problems outlined above, various model surfaces for microbial colonization have been developed that can be worn in the mouth by volunteers. These surfaces can be removed from the mouth to facilitate sampling. The microbiology of fissure plaque has been studied using artificial or natural fissures mounted in a crown or in an occlusal filling. Removable pieces of enamel, denture acrylic, Mylar foil, adhesive tape, spray plast or epoxy resin have been placed on natural teeth or dentures in a desired position and have been used for studies of colonization or of the structural development of dental plaque. Removable appliances have the additional advantage that experiments can be performed on the surfaces when out of the mouth that would not be permitted on natural teeth, e.g. the effect of regular sugar applications, or the evaluation of novel antimicrobial agents.

DISTRIBUTION OF THE RESIDENT ORAL MICROFLORA

The populations making up the resident microbial community of the oral cavity are not found with equal frequency throughout the mouth. Many species are associated with specific oral surfaces. It has been emphasized that the mouth has two distinct types of surface for colonization – hard tissues (teeth) and soft mucosal surfaces. These latter tissues are not homogeneous; the palate contains keratinized surfaces while the cheek is composed entirely of unkeratinized cells. In contrast, the tongue is a specialized structure having a highly papillated surface covered with sensory cells such as taste buds. When these factors are considered along with known differences in the local environment at particular sites, it is not surprising that variations occur in the microbial community at each of these habitats.

In the following sections the predominant microflora from several different sites in the oral cavity will be compared. Little information is available concerning mucosal surfaces; because of its association with disease, most attention has been focused on the microbial composition of dental plaque.

Lips and palate

There is little information about the microbial community of the lips and palate. The lips form the border between the skin microflora, which consists predominantly of staphylococci, micrococci and Gram-positive rods (e.g. *Corynebacterium* spp.), while that of the mouth contains many Gram-negative species and few of the organisms commonly found on the skin surface. A study of this transition area would be of great interest in that it might indicate how the microflora at this site copes with the constant pressure of contamination from two contrasting eco-systems. It is likely that facultatively anaerobic streptococci comprise a large part of the microflora on the lips. *Veillonella* and *Neisseria* have been found, but only in very low numbers (0.38% and <0.05% of the total cultivable microflora, respectively). *Candida albicans* can colonize damaged lip mucosal surfaces in the corners of the mouth ('angular cheilitis'; Chapter 9). *S. vestibularis* is recovered most commonly from the 'gutter' between the lower lip and the gums, and occasionally black-pigmented anaerobes and fusobacteria have been detected.

The microflora of the normal palate can show large variations between subjects, not only in the total colony forming units removed (which may reflect differences in the area sampled and the success in removing organisms) but also in the proportions of the individual species. The majority of the bacteria are streptococci and actinomyces; veillonellae and Gram-negative anaerobes are also regularly recovered but at lower levels (Table 4.2). Haemophili have also been found when the appropriate selective media have been used. Candida are not regularly isolated from the normal palate although their prevalence does increase if dentures are worn. The mucosa of the palate can become infected with *C. albicans* (denture stomatitis); oral candidosis will be considered in more detail in Chapter 9.

Table 4.2 Predominant microflora of the healthy human palatal mucosa

Micro-organism	Percentage of the total cultivable microflora	Percentage isolation frequency
Streptococcus	52	100
Actinomyces	15	100
Lactobacillus	1	87
Neisseria	2	93
Veillonella	1	100
'Bacteroides'	4	100
Candida	+*	7

* Present, but in numbers too low to count.

Cheek

The predominant populations of bacteria isolated from the cheek (buccal mucosa) are shown in Table 4.3, while Table 4.4 gives the relative proportions of streptococci at this site as reported in a separate study. In both studies, *S. sanguis* and *S. mitis* (biovar 1) were among the predominant species (Tables 4.3 and 4.4). Significantly, the majority of the streptococcal strains isolated were IgA$_1$ protease producers. As with other mucosal surfaces, obligate anaerobes are not regularly isolated and when present they do not constitute a large percentage of the microflora. Haemophili (especially *H. parainfluenzae*) are commonly isolated in moderately-high numbers. Spirochaetes and other motile organisms have been observed occasionally by phase-contrast microscopy attached to the buccal mucosa. The concentration of micro-organisms on the cheek epithelium is similar to that of the palate (5–25 bacteria per cheek epithelial cell). *Simonsiella* spp. are isolated primarily from the cheek cells of man and animals.

Tongue

The dorsum of the tongue with its highly papillated surface has a large surface area and therefore supports a higher bacterial density than other oral mucosal surfaces. Microscopic studies have indicated a mean concentration of 100 bacteria per tongue epithelial cell, while cultural studies have demonstrated the presence of a relatively diverse

Table 4.3 Proportions of some bacterial populations at different sites in the normal oral cavity

Bacterium	Saliva	Buccal mucosa	Tongue dorsum	Supra-gingival plaque
S. sanguis	1	6	+*	7
S. salivarius	3	3	6	2
S. mitis	21	29	33	23
mutans streptococci	4	0	3	5
A. viscosus	1	10	3	8
A. naeslundii	2	1	5	5
A. odontolyticus	2	1	7	13
Haemophilus spp.	4	7	15	7
Capnocytophaga spp.	<1	<1	1	<1
Fusobacterium spp.	1	<1	<1	<1
Black-pigmented anaerobes	<1	<1	1	+*

* Detected on occasions.

Table 4.4 Relative proportions of streptococci at different sites in the healthy mouth

	Cheek	Tongue	Pharynx	Supra-gingival plaque	Sub-gingival plaque
S. sanguis	49	−*	2	13	2
S. gordonii	2	+†	4	9	+
S. oralis	+	−	+	5	+
S. mitis biovar 1	22	13	21	10	5
S. mitis biovar 2	+	52	17	6	2
S. salivarus	2	17	30	+	−
S. vestibularis	2	+	4	6	−
S. anginosus	4	12	14	16	85
S. mutans	−	−	−	1	−

Results are expressed as percentages (median values) of the total streptococcal count
* Not detected
† Recovered on occasion.

microflora with several obligately anaerobic species present. The relative proportions of the resident microflora are shown in Table 4.3, with the most prevalent facultatively anaerobic streptococci listed in Table 4.4. Streptococci are the most numerous group of bacteria (approx 40% of the total cultivable microflora) with *S. salivarius, S. anginosus* and *S. mitis* predominating.

Anaerobic streptococci (*Peptostreptococcus* spp.) have also been isolated in some studies while *Stomatococcus mucilagenosus* is found almost exclusively on the tongue. Other major groups of bacteria (and their proportions) include *Veillonella* spp. (16%), Gram-positive rods (16%) of which *A. naeslundii* and *A. odontolyticus* are common, and haemophili (15%). Both pigmenting (*P. intermedia, P. melaninogenica*) and non-pigmenting anaerobes can be recovered from the tongue and this site is regarded as a potential reservoir (along with the tonsils) for some of the organisms implicated in periodontal diseases. Other organisms including lactobacilli, yeasts, fusobacteria, spirochaetes and other motile organisms have been found in low numbers (<1% of the total microflora).

Saliva

Although saliva contains up to 10^8 micro-organisms ml^{-1} it is not considered to have its own microflora. The normal rate of swallowing ensures that bacteria cannot be maintained in the mouth by multiplication in saliva. It is believed that the organisms found are derived from other surfaces as a result of the removal forces (saliva and GCF flow, chewing, oral hygiene) operating in the mouth. A comparison of the composition of saliva with that of other surfaces (Table 4.3) has led to the conclusion that the tongue is the major source of salivary bacteria. Attempts have been made to use the microbial composition of saliva (in particular, the level of mutans streptococci and/or lactobacilli) as an indicator of the caries susceptibility of a mouth, and commercially-available kits for their culture are available. Those with high counts are considered to be 'at-risk' individuals, and can be targeted for intense oral hygiene and dietary counselling.

Teeth

The microbial community associated with teeth is referred to as dental plaque. Its composition varies on different surfaces due to the local environmental conditions (Chapter 2). For these reasons, plaque is described on the basis of the sampling site by terms such as smooth surface, approximal, fissure, or gingival plaque (Figure 2.2). Similarly, samples taken from above the level of the gum margin are given the general name supra-gingival plaque, while those from below the gum margin are described as sub-gingival plaque. The detailed composition of dental plaque from these sites will be given in Chapter 5. As teeth are non-shedding surfaces, the highest concentrations of micro-organisms are found in stagnant sites which afford protection from removal forces.

Table 4.3 shows that unlike some oral surfaces, Gram-positive filaments (mainly *Actinomyces* species) are a major group of bacteria in plaque. Mutans streptococci and members of the *S. oralis* and *'S. milleri'*-groups are generally found in highest numbers on teeth (Table 4.4), and these organisms have a strong affinity for hard surfaces. Indeed, these species do not usually appear in the mouth until after tooth eruption. In addition, *S. sanguis* can be isolated from many of the exposed surfaces of teeth such as smooth surfaces. In contrast to mucosal surfaces, *S. salivarius* is only a minor component of dental plaque. Haemophili are present in moderate numbers, although the individual species can differ from those found on other oral surfaces. Obligate anaerobes are found in high numbers particularly in the gingival crevice, and oral spirochaetes are almost uniquely associated with this region. Thus, the composition of dental plaque differs both qualitatively and quantitatively from the communities of other oral surfaces.

Summary

Oral populations are not distributed evenly in the mouth. In particular, large differences occur in the prevalence of individual streptococcal species at particular oral sites (Tables 4.3 and 4.4). This results in each habitat having a characteristic microflora.

FACTORS INFLUENCING THE DISTRIBUTION OF ORAL MICRO-ORGANISMS

Anaerobiosis

Large variations in redox potential (Eh) occur around the mouth. In Chapter 2, the sites with the lowest Eh were stated to be those associated with the teeth, such as approximal areas and the gingival crevice, and these are the areas that also have the highest concentration of obligate anaerobes. Anaerobes are found primarily at stagnant sites where plaque accumulates and gradients in oxygen tension and Eh can develop. The gingival crevice harbours spirochaetes and the largest numbers of obligately anaerobic Gram-negative rods and filaments. In contrast, the exposed surfaces of the teeth and the epithelial tissues have much lower levels of these anaerobes. Thus, the distribution of the anaerobes follows that of the redox potential at sites around the mouth.

Nutrition

Attempts have been made to relate the distribution of some oral bacterial populations to nutrient availability. As stated earlier, some of the bacteria isolated from plaque are nutritionally-fastidious and are difficult to culture in the laboratory unless special precautions are made, and yet all of their growth requirements must be met by their natural habitat. Thus, black-pigmented anaerobes require haemin, and they generally obtain this cofactor from the degradation of host macromolecules such as haemoglobin, haemopexin and haptoglobin. Likewise, spirochaetes are reported to require α–2-globulin and, as both this nutrient and haeme-containing proteins are found mainly in gingival crevicular fluid, it is not surprising that both of these groups of bacteria are found in the highest numbers in sub-gingival plaque. The increased proportions of many asaccharolytic but proteolytic bacteria in sub-gingival plaque during various periodontal diseases is considered, in part, due to the provision of novel nutrients (proteins, glycoproteins) by GCF (Chapter 7). The presence of other species can be attributed not to the presence of host-derived nutrients but by the provision of nutrients by other oral bacteria themselves. Thus, some species use end products of metabolism of other organisms (e.g. lactate, succinate) while for others, key nutrients such as peptides or amino acids are generated during the metabolism of host molecules by other bacteria. Indeed, the degradation of many macromolecules is now being shown to involve or to be enhanced by the metabolic co-operation of mixtures (consortia) of bacteria.

Adherence

For successful colonization, populations must first adhere to, or be retained at, a surface and then be able to multiply. Several factors including saliva and crevicular fluid flow, chewing and oral hygiene procedures serve, in part, to remove micro-organisms from oral surfaces. In order to overcome these removal forces, populations either seek out habitats that offer ecological refuge from the environment, or become specifically adapted to such adverse conditions. The former microbial approach is non-specific and possibly relates to weakly-adhering strains. This has been referred to as retention and can be distinguished from adherence, which involves specific physico-chemical mechanisms.

Several oral populations have been compared in their ability to adhere *in vitro* to teeth and to epithelial cells derived from scrapings of the human tongue and cheek. Certain species were found to adhere consistently in higher numbers than others to particular surfaces so that gradings for adherence (low, moderate or high) could be given (Table 4.5). These *in vitro* experiments were repeated *in vivo* using streptomycin-resistant strains of mutans streptococci, *S. sanguis* and *S. salivarius*. After standardized suspensions of a mixture of the three species were rinsed around the mouths of volunteers, the adherence was graded as before. Similar patterns were found when the *in vitro* and *in vivo* results were compared. These experimentally-observed patterns correlated closely with the natural prevalence of each species at these sites (Table 4.5). Bacteria adhere with considerable specificity not only to certain tissues, but also to that particular tissue of their particular animal host. Thus, the distribution of many oral populations appears to be related to their ability to adhere to specific surfaces.

The paucity of some populations on various oral surfaces has been attributed to their poor ability to adhere. However, many poorly-adhering bacteria such as *Veillonella* are present in high numbers at stagnant sites in the mouth of man and a variety of animals. These organisms are usually found associated with the protected areas of the mouth, for example, pits and fissures, approximal areas and the gingival crevice, and are not normally isolated in high numbers from the more exposed oral surfaces such as the cheek or the smooth surface of the tooth. Thus, it is likely that these populations are unable to adhere strongly enough to withstand the oral removal forces and are retained in the mouth only by having gained refuge at such stagnant sites. When these populations are found in exposed regions they may be protected from the environment by extracellular polysaccharides produced by strongly adherent species such as the oral streptococci, or by secreted enzymes (glucosyltransferases) that become adsorbed to their cell surface ('surrogate enzymes'). Alternatively, they may find protection in the small pits and defects that occur commonly on the enamel surface (Figure 4.3). Only in the absence of firm removal forces can the weak affinity of these organisms for an oral surface prove sufficient for anchorage. The essential differences between adherence and retention are illustrated in Figure 4.4.

Summary

The distribution of organisms in the mouth is related to the Eh and nutrient availability at individual sites, and also to the strength of adherence between an organism and a surface.

Table 4.5 A comparison of the ability of some bacteria to adhere to oral surfaces, with their relative proportions at these sites

Population	Experimentally-observed adherence			Relative indigenous proportions		
	Tooth	Tongue	Cheek	Tooth	Tongue	Cheek
mutans streptococci	low–high*	low	low	low–high*	low	low
S. sanguis	high	moderate	moderate	high	moderate	moderate
S. mitis	high	moderate	high	high	moderate	high
S. salivarius	low	high	moderate	low	high	moderate
Lactobacillus	low	low	low	low	low	low
Veillonella	low	high	low	low†	high	low
Neisseria	low	low	low	low	low	low

* High under the influence of dietary sucrose
† Veillonella can be found in high numbers in dental plaque.

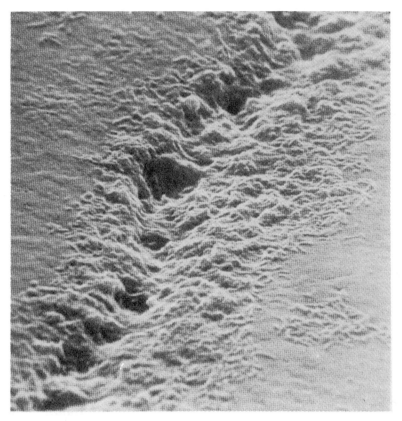

Figure 4.3 Oral micro-organisms seeking ecological refuge in a defect in the enamel surface (magnification approx. × 2400). (Courtesy of C. A. Saxton)

FACTORS INVOLVED IN ADHERENCE

Microbial adherence is one of the most active fields of research not only in oral ecology but also in other aspects of microbiology. As adherence is essential for colonization by pathogenic as well as by commensal micro-organisms, any approach that can successfully interfere with these processes could have far-reaching implications.

The first stage of adherence involves the initial interaction between micro-organism and substrate. It involves the external surfaces of both organism and substrate and will be influenced by the suspending medium. To understand microbial attachment in the mouth it is essential to determine the interactive polymers on the surface of the oral organisms and on the colonizable substrates (teeth, epithelial surfaces). In addition, the influence of saliva as a suspending fluid must be under-

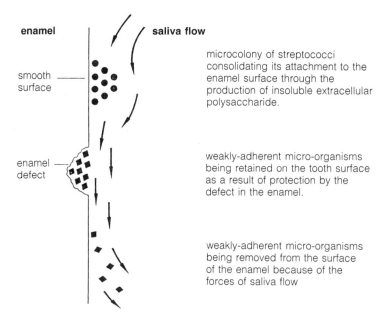

enamel

smooth
surface

enamel
defect

saliva flow

microcolony of streptococci
consolidating its attachment to the
enamel surface through the
production of insoluble extracellular
polysaccharide.

weakly-adherent micro-organisms
being retained on the tooth surface
as a result of protection by the
defect in the enamel.

weakly-adherent micro-organisms
being removed from the surface
of the enamel because of the
forces of saliva flow

Figure 4.4 Diagram to distinguish adherence from retention.

stood. The complexity of these interacting components makes this a particularly difficult area for study. In general, the nomenclature used is for the bacterial components which function in adherence to be termed **adhesins** while the host-derived factors are called **ligands**. It will become clear in the following sections that a bacterial cell surface contains several types of adhesin while the host surface can express multiple ligands. Consequently, the molecular interactions by which a given strain attaches to ligands on dissimilar surfaces (e.g. enamel, buccal mucosa, etc) may well be different. The polymers of host and bacterial origin that are considered to be important in adherence are discussed in the following sections.

Host ligands

Mammalian epithelial cells, and especially buccal epithelium, have sialic acid exposed on their surfaces which can interact with receptors on bacteria e.g. *S. mitis*. If the sialic acid residue is removed by neuraminidase, then another receptor (a galactosyl ligand) is exposed which is recognized by *Actinomyces* spp., and Gram-negative bacteria including *F. nucleatum*, *Prevotella intermedia* and *E. corrodens*. Collagen fibres, which are major structural components of connective tissue, can also act as receptors for certain mutans streptococci (*S. cricetus*, *S. rattus*) and *Porphyromonas gingivalis*.

The major groups of host receptors are found in saliva; these receptors, when adsorbed onto oral mucosal and enamel surfaces, will influence bacterial attachment. The greatest amount of research has been directed towards the receptors found on the enamel surface in what is termed the acquired pellicle. This pellicle is generally <1 μm thick and is formed by the selective adsorption of components, mainly from saliva but also from GCF. Pellicles form on all oral surfaces (hard and soft) and are not identical; the components that adsorb to cementum are not the same as enamel, and both will differ from those which form on the oral mucosa. These differences are sometimes acknowledged by the use of different terminologies, e.g., the acquired enamel pellicle or the acquired cementum pellicle, while the pellicle that forms on epithelial surfaces is referred to as the mucus coat. Pellicle forms as soon as a clean surface is exposed to saliva; it takes 90–120 minutes for the adsorption of molecules to reach a plateau and cease. Pellicles contain proteins, lipids and glycolipids, but little is known of the conformational state of the adsorbed molecules. Once formed, the composition and structure of pellicles will change and be modified.

Within the enamel pellicle, acidic proline-rich proteins and statherin promote the adherence of *A. viscosus*, some *S. mutans* strains and black-pigmented anaerobes. However, as hydroxyapatite is amphoteric due to the presence of phosphate and calcium atoms, both basic and acidic compounds are adsorbed; some sulphur-containing amino acids are also found. Immunological probes have provided evidence for the presence of amylase, lysozyme, albumin and immunoglobulins in the acquired pellicle. Likewise, some bacterial components have also been recognized, including glucosyltransferases (GTFs) and glucans. Many of these adsorbed compounds can act as ligands for oral bacteria. GTFs are enzymes that synthesize glucans from sucrose. The adsorbed enzymes in the pellicle are still able to function and the glucans produced may bind to receptors on bacteria, such as *S. mutans*, increasing their ability to colonize.

Bacterial adhesins

Many bacterial adhesins are lectins (carbohydrate-binding proteins) which bind to carbohydrate receptors on a surface. Often these adhesins are associated with surface structures termed fibrils or fimbriae. Fibrils can be distinguished from fimbriae in that they clump together and have no measurable width while fimbriae have a measurable width (3–14 nm) and a variable length up to 20 μm. Some cells possess both

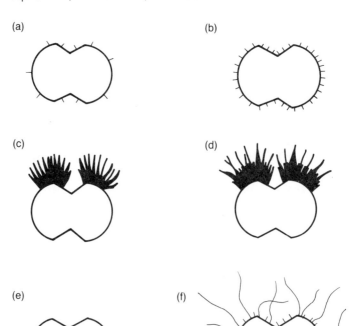

Figure 4.5 Diagrammatic representation of the surface structures found on streptococci belonging to the *S. oralis*-group. Cells possess fibrils of different lengths at different densities. Patterns (a) and (b) are found most commonly; patterns (c) and (d) are found on 'tufted' streptococci, and these strains are involved in co-aggregation with *C. matruchotii* to form 'corn cob' structures (Chapter 5).

fibrils and fimbriae, and strains can have different types of fibrils or fimbriae (Table 4.6). For example, several patterns of surface structure were observed when strains of the *S. oralis*-group were examined (Figure 4.5). These fibrils could be of different length (long fibrils = 160 nm, short fibrils = 50 nm), at different densities, and with possibly different functions. Strains with fibrils arranged in tufts consistently lacked the ability to co-aggregate with *Actinomyces* spp. and came from a variety of habitats other than supra-gingival plaque. Similarly, *S. salivarius* strains have been found to carry a complex fibrillar mosaic comprising four different classes of fibril, each with a specific length. The 91 nm fibrils are responsible for co-aggregation with *V. parvula* while the 73 nm fibrils are involved with adhesion to buccal epithelial cells. The longest (178 nm) and shortest (63 nm) fibrils have yet to be ascribed a function. Underneath these fibrils, but outside the cell wall, is a dense ruthenium red staining layer, the function of which is also unknown. The presence of fibrils is not limited to Gram-positive oral bacteria. Strains of *P. intermedia* carry peritrichous fibrils, the morphology, cellular density and length of which vary sharply among strains.

Table 4.6 The types of surface structures found on oral bacteria

| Species | Surface structure | | | |
	Fibrils	Tufts of fibrils	Fimbriae	No structure
S. mutans	Rare	–	–	+
S. oralis-group	+	+	Rare	–
S. anginosus	–	+	+	+
S. salivarius	+	–	+	–
A. viscosus	–	–	+	–
A. naeslundii	–	–	+	–
P. intermedia	+	–	–	–
P. melaninogenica	+	–	–	–
P. loescheii	–	+*	–	–
P. buccae	+	–	–	–
P. oralis	+	–	–	–
P. gingivalis	–	–	+	+

* Structures were originally described as fimbriae but would now be classified as fibrils.

A. *viscosus* cells possess two types of fimbriae. Type 1 fimbriae mediate the binding of cells to salivary pellicle on enamel whereas type 2 fimbriae are associated with a galactosyl-binding lectin which mediates attachment of *A. viscosus* to epithelial cells and to polymorphs, or to other bacteria (co-aggregation). The type 1 fimbriae bind to adsorbed proline-rich proteins and to statherin. Lactose-inhibitable coaggregation has also been observed between *Prevotella* spp. and *A. israelii* or *S. sanguis*. This coaggregation has been studied in *P. loescheii* and has been shown to be mediated by fimbria-associated proteins; a 75- and 43- kDa polypeptide being responsible for the recognition of *S. sanguis* and *A. israelii*, respectively.

Other bacterial adhesins include GTF, which are found on the surface of several oral streptococci. These GTFs can interact with receptors in pellicle such as blood group reactive proteins or adsorbed dextrans and glucans. The latter can also react with glucan-binding proteins present on certain streptococci. The polysaccharides synthesized by GTFs help consolidate bacterial attachment to hard surfaces in the mouth and contribute to the plaque matrix (Chapter 5). Many oral Gram-positive bacteria are negatively charged due to the penetration of the cell wall by lipoteichoic acid (LTA). These anionic polymers are composed of sugar phosphates, usually glycerol and ribitol phosphate. LTA has also been shown to interact with blood group reactive substances in pellicle. As antibodies are found in the acquired pellicle, a number of antigens may also act indirectly as 'adhesins'. Similarly, lysozyme in the pellicle can also bind to protein receptors on bacteria.

The production of adhesins by bacteria will be influenced by the environment. Changes in the thickness of ruthenium red-staining layers, hydrophobicity, the density of the fibrillar fringe, and the concentration of cell surface proteins can be shown to vary when organisms are grown at different rates.

The acquisition of the resident oral microflora therefore involves the interaction between adhesins on the microbial cell surface and ligands on the host surface. Because of the specificity of these interactions, the number of cells initially adhering to mucosal cells is small; regular desquamation will also ensure that the microbial load is light at these sites. In contrast, a wider range of ligands are adsorbed onto teeth (pellicle), thereby increasing the potential for diversity in terms of microbial colonization. Furthermore, the fact that teeth are non-shedding surfaces means that large accumulations of micro-organisms can develop. These accumulations, or biofilms, are termed dental plaque; the properties and composition of dental plaque will be discussed in Chapter 5.

FUNCTIONS OF THE CLIMAX COMMUNITY: COLONIZATION RESISTANCE

One of the main beneficial functions of the resident microflora at any site is to prevent colonization by exogenous organisms; such organisms are often pathogenic for the host. This property of exclusion has been termed colonization resistance, and the microbial factors that contribute to this property are listed in Table 4.7. Members of the resident oral microflora may exclude exogenous organisms by saturating the available ligands and receptors that are available for attachment. Likewise, the indigenous species may be more competitive for the natural substrates present in the mouth so that invading organisms cannot flourish. The metabolism of the resident microflora can also make conditions unsuitable for colonization by other organisms; examples may include changes in local pH or Eh. Another mechanism is the production of inhibitors, of which there are many examples, including the production of hydrogen peroxide by members of the *S. oralis*-group, bacteriocin production by a range of Gram-positive bacteria, especially streptococci, and the formation of acidic end-products of metabolism. It has been proposed that strains of *S. salivarius* that produce an inhibitor (termed enocin), and which is active against Lancefield Group A streptococci (*S. pyogenes*), can prevent colonization by this pathogen on mucosal surfaces.

Table 4.7 Functions of the resident oral microflora that may contribute to colonization resistance

Function
Competition for receptors/ligands for adhesion
Competition for essential endogenous nutrients and cofactors
Creation of micro-environments that discourage the growth of exogenous species
Production of inhibitory substances

Attempts have been made to enhance the colonization resistance of the resident oral microflora. As the mother is the major source of bacteria (including cariogenic species) in the infant, levels of mutans streptococci have been suppressed in expectant mothers by professional oral hygiene, and dietary counselling with, if necessary, treatment with chlorhexidine or fluoride (Chapter 6). The result is that the natural microflora of the infant increases in complexity in the absence of mutans streptococci. The subsequent colonization by these cariogenic streptococci is therefore delayed, as is the average time for the first caries lesion to form. Other approaches that are still at the laboratory stage are to use pre-emptive colonization with either low virulence mutants of *S. mutans* (e.g. strains deficient in glucosyltransferases, intracellular polysaccharide production, or lactate dehydrogenase activity) or with organisms that are more competitive than wild-type *S. mutans* strains. For example, a strain of *S. salivarius* (strain TOVE) has been shown to displace virulent strains of *S. mutans* from teeth in experimental animal studies while bacteriocin-producing but avirulent mutants of *S. mutans* have been shown to be able to colonize human teeth in volunteers. Both strains could be candidates for possible effector strains for this type of replacement therapy.

Host factors also play a role in colonization resistance. The immune and innate host defences will help exclude invading organisms. However, colonization resistance can also be impaired by factors that compromise the integrity of the host defences or perturb the resident microflora. Classical examples would be the long-term use of broad spectrum antibiotics or of cytotoxic therapy, but other more subtle mechanisms can apply. Fibronectin has been shown to prevent adherence of *Pseudomonas aeruginosa* to buccal epithelial cells. Levels of fibronectin in seriously ill adults and in infants are lower than those in healthy adults and may account for the higher rates of colonization by Gram-negative bacilli in these subjects.

METABOLISM OF ORAL BACTERIA

The persistence of the resident oral microflora is dependent on their ability to obtain nutrients and grow in the mouth. Nutrients are derived mainly from the metabolism of endogenous substrates present in saliva and GCF. Superimposed on these components are exogenous nutrients which are supplied intermittently via the diet; the most significant of these are dietary carbohydrates and casein. The concentration of nutrients will affect the growth rate and physiology of the microflora, as will any changes in pH resulting from microbial metabolism. The fluctuating conditions of nutrient supply and environmental change requires the microflora to possess biochemical flexibility. Their pattern of metabolism is intimately related to whether the resident microflora enjoys a pathogenic or commensal relationship with the host.

Carbohydrate metabolism

Most attention has been paid to the metabolism of carbohydrates because of the relationship between dietary sugars, low pH, and dental caries (Chapters 2 and 6). The metabolic fate of dietary carbohydrates is illustrated in simplified form in Figure 4.6. Starches, which contain mixtures of amylose and amylopectin, can be broken into their constituent sugars by amylases of salivary and bacterial origin. Some streptococci (*S. gordonii*, *S. mitis*) are able to bind salivary amylase, which might

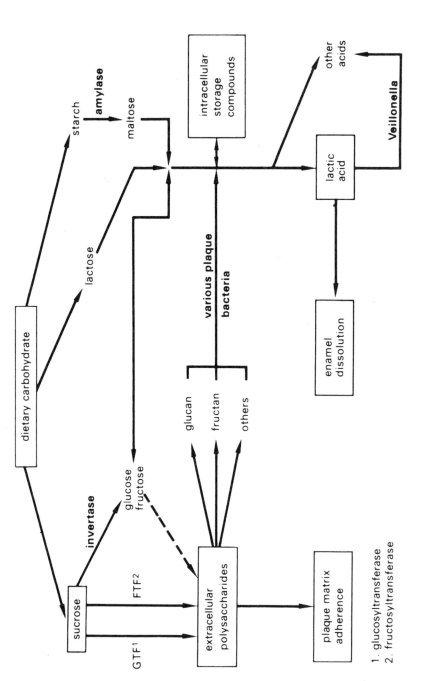

Figure 4.6 Simplified diagram to show the metabolic fate of dietary carbohydrates.

provide additional metabolic capability. *S. mutans* possesses a spectrum of enzymes with the potential to attack the various types of linkage found in dietary starches, although perhaps significantly from a caries standpoint, it does not make acid from starch. These enzymes include ones capable of debranching and degrading amylose and amylopectin, as well as an amylase and an extracellular endodextranase.

Milk is the major source of lactose in the diet, while sucrose occupies a key position in bacterial metabolism in the oral cavity. Sucrose is the most widely used sweetening agent and, in many industrialized societies, consumption is approximately 50 kg/person/year. Sucrose can be:

1. Broken down by extracellular bacterial 'invertase'-like activity and the resultant glucose and fructose molecules taken up directly by bacteria.
2. Transported intact as the disaccharide or disaccharide phosphate, and cleaved inside the cell by an intracellular invertase or a sucrose phosphate hydrolase.
3. Utilized extracellularly by glycosyltransferases. Glucosyltransferases (GTF) produce both soluble and insoluble glucans (with a release of fructose) which are important in plaque formation and in the consolidation of bacterial attachment to teeth. Fructosyltransferases (FTF) produce fructans (and liberate glucose) which are frequently labile and can be used by other plaque organisms.

Some aspects of the metabolism of carbohydrates will be considered in more detail in the following sections.

Sugar transport and acid production

All substrates have to be transported across the cytoplasmic membrane and into the bacterial cell if they are to be of value for biomass production or as an energy source. Sugar transport by a phosphoenolpyruvate-mediated phosphotransferase (PEP-PTS) system has been reported for oral strains of the genera *Streptococcus*, *Actinomyces*, and *Lactobacillus*. The PEP-PTS is a carrier-mediated, group translocating process in which two non-specific, general cytoplasmic proteins (HPr and enzyme I) transfer high energy phosphate from PEP to the sugar moiety via a sugar-specific, membrane-bound enzyme II, leading to the formation of sugar phosphate and pyruvate (Figure 4.7). The transport of particular sugars in some bacteria also involves an enzyme III, which can be cytoplasmic or bound to the cytoplasmic membrane. Most, if not all,

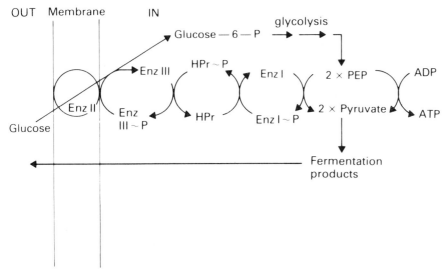

Figure 4.7 Diagrammatic representation of a phosphoenolpyruvate (PEP)-mediated glucose phosphotransferase (PTS) system. Enz I, II and III represent enzymes I, II and III, respectively; HPr is a heat stable protein. Other carbohydrates are also transported by this system.

mono- and disaccharides are transported into bacteria belonging to the genera listed above by this system. Some of the individual components have been isolated and characterized; some sugars appear to possess more than one enzyme II. Generally, the PEP-PTS is constitutive, although the activity of the system is modulated by environmental conditions. It is most active when sugars are in low (growth-limiting) concentrations, and is repressed under sugar excess conditions. However, lactose can induce the lactose-PTS uptake system in oral streptococci. For example, when *S. mutans* was grown on lactose there was co-induction of the lactose-PTS and phospho-β-galactosidase, the role of the latter enzyme being to cleave intracellular lactose-phosphate to galactose-6-phosphate and glucose. Similarly, growth of *S. mutans* on mannitol or sorbitol leads to the induction of distinct PTS systems for these sugar alcohols, although they are rapidly repressed by glucose.

The PTS systems are also susceptible to environmental pH, and are repressed under acidic conditions. Paradoxically, acid production and growth of mutans streptococci is enhanced at low pH, and this finding led to the search for alternative sugar uptake systems. Studies of *S. mutans* in which PTS-mediated sugar transport was abolished by sub-lethal concentrations of chlorhexidine, and of a mutant defective in glucose-PTS activity, showed that cells still retained some residual glycolytic activity. This activity could be abolished by inhibitors of membrane activity (uncouplers, ionophores), thereby providing strong evidence for an uptake system driven by the energy associated with the bacterial membrane (protonmotive force, pmf). This alternative system transports carbohydrates as the free sugar, which is then phosphorylated intracellularly in order to become available metabolically to the cell. In *S. mutans*, this alternative system operates maximally at around pH 5.5, and allows the cell to take advantage of the transmembrane pH gradient generated when the extracellular pH is low compared to the intracellular pH.

Kinetic studies have shown that the pmf-driven system is the low affinity system in *S. mutans*, whereas the PTS is a high affinity scavenger system operating principally under substrate-limited conditions. In this way, these organisms are able to take advantage of the fluctuating 'feast-and-famine' conditions in the mouth in terms of the availability of dietary sugars. Mutans streptococci and lactobacilli are also able to cope better than other species with the low pH so-generated from carbohydrate metabolism. This metabolic flexibility in terms of coping with oscillating extremes of sugar concentration and pH contribute to the ability of mutans streptococci to predominate during the development of a caries lesion (Chapter 6).

One group failed to detect the presence of a pmf-mediated sugar transport system and, on the basis of their results, proposed that carbohydrates could enter the cell by passive diffusion on a concentration gradient. Others have found evidence for another type of transport system for melibiose (a disaccharide with the composition galactose-glucose) based on a comparison of nucleotide sequences of the genome of *S. mutans* with those with a known function in other bacteria. *S. mutans* was found to contain similar DNA base pair sequences to those of periplasmic transport genes of Gram-negative species. If confirmed, this would be the first finding of such a transport system in a Gram-positive organism.

In addition to dietary sugars, the resident oral microflora also obtain carbohydrates for biomass and energy from the catabolism of host glycoproteins present in saliva (e.g. mucins) and GCF (e.g. transferrin). Bacteria produce a range of glycosidases that can remove sugars sequentially from the oligosaccharide side chains of these glycoproteins. As no single species possesses the full enzyme complement, bacteria interact synergistically to fully degrade these molecules (Chapter 5). Acid production from these glycoproteins is slow compared to that from exogenous sugars, and would not cause significant enamel demineralization. Once sugars have been transported into the bacterial cell, they can be used either in anabolic pathways to generate biomass, or they can be

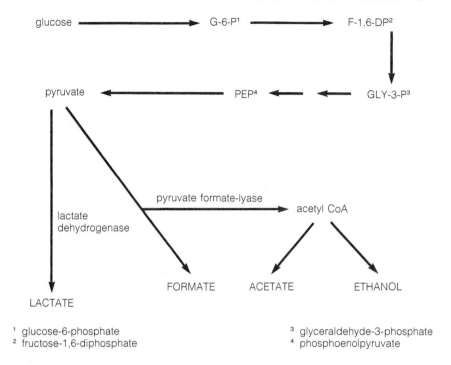

glucose ⟶ G-6-P¹ ⟶ F-1,6-DP²

pyruvate ⟵ PEP⁴ ⟵ ⟵ GLY-3-P³

pyruvate formate-lyase → acetyl CoA

lactate dehydrogenase

FORMATE   ACETATE   ETHANOL

LACTATE

¹ glucose-6-phosphate   ³ glyceraldehyde-3-phosphate
² fructose-1,6-diphosphate   ⁴ phosphoenolpyruvate

Figure 4.8 Formation of end products of metabolism by mutans streptococci.

broken down to organic acids (which are excreted) to generate energy. Acid production has been studied intensively because of its role in the demineralization of enamel. Bacteria catabolize sugars by glycolysis to pyruvate; the fate of pyruvate will depend on the particular organism and the availability of oxygen. Most oral bacteria metabolize pyruvate anaerobically to organic acids, the pattern of which can sometimes be used in the identification of particular genera (Chapter 3). Oral strepto-cocci convert pyruvate to lactate by lactate dehydrogenase when sugars are in excess, while formate, acetate and ethanol are the products of metabolism of mutans streptococci and *S. sanguis* (but not *S. salivarius*) under carbohydrate limitation (Figure 4.8). Other bacterial genera prod-uce acetate, butyrate, propionate, and formate as primary products of metabolism. Different species produce acid at different rates, and vary in the terminal pH reached and in their ability to survive under such conditions. Mutans streptococci produce acid at the fastest rates while lactobacilli generate the lowest environmental pH; both groups are also aciduric and can tolerate conditions of acidity that most other oral bacteria would find inhibitory or even lethal (Figure 2.5 and Table 6.8).

The rate and extent of acid production varies with other environmental conditions; for example, a high potassium/sodium ratio, as found in plaque fluid, enhances glycolysis. The properties of micro-organisms are also known to vary when grown on a surface. Acid production from glucose by mutans streptococci was enhanced when cells were attached to hydroxyapatite beads, although this stimulation was not found with all species of oral streptococci. The metabolism of substrates by oral bacteria growing in a biofilm is an area that has been largely ignored and yet is one that is highly relevant to the mouth in health and disease.

Variations are found in the profiles of acids found in plaque at different times of the day. Acetic, succinic, propionic, valeric, caproic and butyric acids were found in human and monkey plaque sampled after overnight fasting. These profiles reflect heterofermentation and amino acid catabolism. Following exposure to sucrose, the concentration of volatile acids fall while lactic acid becomes the predominant fermentation product. Such a switch in metabolism will encourage demineralization.

Polysaccharide production

Bacteria in the mouth are subjected to continual cycles of 'feast and famine' with respect to dietary carbohydrates. As a consequence, the resident microflora has developed strategies to store these carbohydrates during their brief exposure to these energy sources. The most common strategy is to store these carbohydrates as intracellular polysaccharides (IPS), and many species of oral streptococci can synthesize polymers that resemble glycogen (1,4–α-glucan), although other polymers might also be formed.

In a rat test system, the virulence of IPS-defective mutants of *S. mutans* was compared with that of the parent strain. Although the mutants were able to colonize and persist on tooth surfaces, they differed from the parent strains in causing fewer caries lesions both in fissures and on smooth surfaces. This difference in pathogenicity was attributed to the inability of the mutants to produce acid from IPS reserves in the absence of exogenous supplies of carbohydrate. Early epidemiological surveys of humans found an association between the presence of IPS-producing bacteria in plaque and the incidence of dental caries. However, caution should be exercised when generalizing over the role of IPS in virulence since strains of *S. sobrinus* characteristically have low IPS levels and yet have been implicated in the aetiology of human caries (Chapter 6).

It has been generally assumed that glycogen synthesis and degradation in oral streptococci follow the conventional pathway:

$$\text{glucose-6-phosphate} \xrightleftharpoons{\text{phosphoglucomutase}} \text{glucose-1-phosphate}$$

$$\text{glucose-1-phosphate} + \text{ATP} \xrightleftharpoons{\text{ADP-glucose synthase}} \text{ADP-glucose} + \text{PP}_i$$

$$\text{ADP-glucose} + (\text{glucose})_n \xrightleftharpoons{\text{ADP-glucose transferase}} (\text{glucose})_{n+1} + \text{ADP}$$

$$(\text{glucose})_{n+1} + \text{P}_i \xrightleftharpoons{\text{glycogen phosphorylase}} \text{glucose-1-phosphate} + (\text{glucose})_n$$

However, in a survey of a range of oral streptococci, little phospho-glucomutase activity could be detected and a novel first step in glycogen synthesis was discovered. Glucose, transported across the energized

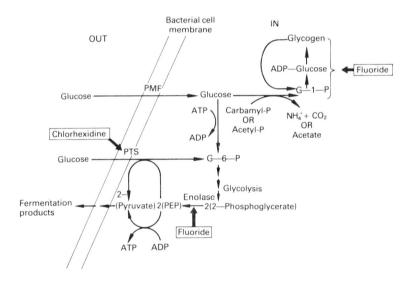

Figure 4.9 Routes of glucose uptake and metabolism in oral strepto-cocci. Possible sites of inhibition by fluoride and chlorhexidine are indicated. G-1-P: glucose-1-phosphate; G-6-P: glucose-6-phosphate; PEP: phosphoenolpy-ruvate; PMF: protonmotive force; PTS: phosphotransferase sugar uptake system.

membrane as free glucose by the pmf-driven uptake system, could be phosphorylated either by ATP to glucose-6-phosphate for glycolysis, or by carbamyl phosphate or acetyl phosphate to glucose-1-phosphate for glycogen synthesis (Figure 4.9). Carbamyl phosphate is generated from the metabolism of arginine, and S. sanguis is able to produce IPS under both carbohydrate-limited and carbohydrate-excess conditions.

Many species of oral bacteria are also able to synthesize extracellular polysaccharides (EPS) from carbohydrates, especially from sucrose (Table 4.8). The polysaccharides can be soluble or insoluble; the former are more labile and can be metabolized by other bacteria while the latter make a major contribution to the structural integrity of dental plaque and can consolidate the attachment of bacteria in plaque. Sucrose has a unique property as a substrate in that the bond between the glucose and fructose moieties has sufficient energy on cleavage to support the synthesis of polysaccharide. The polysaccharides formed are either glucans or fructans, and are synthesized by glucosyltransferases (GTF) and fructosyltransferases (FTF), respectively. The rate of synthesis of EPS is increased in the presence of dextran primer in S. mutans but not in S. sanguis and S. salivarius.

There are a number of enzymes involved in the synthesis of these polysaccharides. GTF-S enzymes synthesize soluble glucan while GTF-I and GTF-SI enzymes are responsible for insoluble glucan formation. The former molecule is a highly branched dextran ($1,6-\alpha-$) with some $1,3-\alpha-$ branch linkages, while the latter is a linear $1,3-\alpha$-glucan, which has been termed mutan. These two classes of enzyme work in concert to form a range of polymers with varying proportions of $1,6-\alpha-$ and $1,3-\alpha-$ linkages. GTF enzymes are secreted and have been found both in pellicle and on the surfaces of unrelated bacteria where they retain biological function. Thus, some species that would appear from laboratory studies to be incapable of making EPS may in fact produce polysaccharides in plaque from 'surrogate enzymes'. GTF that is adsorbed to pellicle can synthesize glucans; these glucans can interact with glucan-binding proteins present on the cell surfaces of oral streptococci and thereby encourage attachment to enamel.

There are also a range of FTF enzymes that produce fructans with different structures. S. mutans produces an unusual water-soluble fructan (inulin) with $2,1-\beta$-linkages which is in contrast to the fructans of other oral bacteria, e.g., S. salivarius, which are levans with a characteristic $2,6-\beta$-linkages (Table 4.8). The FTF of most organisms are secreted but that of S. salivarius is cell-associated. The secretion of GTF and FTF is regulated by the protonmotive force and by the fluidity of the bacterial membrane (although it should be noted that pmf and membrane fluidity are inter-related). The synthesis and activity of these glycosyltransfer-

Table 4.8 Extracellular polysaccharide-producing bacteria found in the oral cavity

Bacterium	Carbohydrate substrate	Type of polymer (with predominant linkages*)
Streptococcus mutans	Sucrose	Water-insoluble glucan (mutan) 1,3-α-; 1,3-α- + 1,6-α-
	Sucrose	Water soluble glucan (dextran) 1,6-α-
	Sucrose	Fructan 2,1-β-
Streptococcus sanguis§/S. gordonii	Sucrose	Water-insoluble glucan 1,3-α- + 1,6-α-
	Sucrose	Water soluble glucan (dextran) 1,6-α-
Steptococcus salivarius‡	Sucrose	Fructan (levan) 2,6-β-
Actinomyces viscosus	Sucrose	Fructan (levan) 2,6-β-
	–†	Heteropolysaccharide (60% N-acetyl glucosamine)
Lactobacillus sp.	–†	Glucan Heteropolysaccharide
Rothia dentocariosa	Glucose	Heteropolysaccharide
	Sucrose	Levan
Stomatococcus mucilagenosus	–†	Heteropolysaccharide (hexoses, hexosamines, amino acids)
Neisseria sp.	Sucrose	Glycogen-like

* Where known
† No specific substrate required
§ Some strains produce a fructan
‡ Some strains produce a glucan.

ases is also regulated by environmental conditions. For example, *S.*

sanguis was believed not to make fructan until cells were grown in media with a high potassium ion concentration. GTF activity of *S. mutans* and *S. gordonii* is also stimulated by potassium, and the significance of these findings is that plaque fluid has a high potassium/sodium ion ratio.

The GTF and FTF enzymes have proved difficult to characterize biohemically, as they are labile and susceptible to rapid degradation by bacterial proteases during purification. Currently, therefore, much effort is being spent on trying to determine the properties and numbers of streptococcal glycosyltransferases using recombinant DNA techniques. It has been found that *S. mutans* and *S. downei* have at least three and four distinct GTF enzymes, respectively, while three FTF enzymes have been recognized in *S. mutans*. It is still to be resolved how these enzymes are regulated and interact to produce extracellular polysaccharide.

Other species that produce EPS are listed in Table 4.7. The heteropolysaccharides are generally complex in composition; for example, the polymer produced by a strain of *A. viscosus* contains N-acetyl glucosamine (62%), galactose (7%), glucose (4%), uronic acids (3%), and small amounts of glycerol, rhamnose, arabinose, and xylose. Some of these homo- and heteropolysaccharides can be metabolized by other oral bacteria. These interactions will be discussed in the next chapter.

Nitrogen metabolism

In contrast to the large amount of data on the metabolism of carbohydrates, much less is known about the metabolism of nitrogenous compounds by oral bacteria. This situation is beginning to change with the appreciation of the role of salivary proteins/glycoproteins as primary nutrient sources in the mouth, and the significance of proteases in the aetiology of periodontal diseases.

Apart from casein, there is little evidence that dietary proteins are utilized to any great extent. Casein can be incorporated into dental plaque and degraded. *S. sanguis* has been shown to have both endo- and exopeptidase activity that can cleave proteins such as casein into a range of peptide fragments. *S. sanguis* can rapidly release arginine from C-terminal peptides, converting the released arginine to energy (and carbamyl phosphate) via the arginine deiminase pathway. Other oral bacteria have different preferences for individual amino acids; for example, *S. mutans* and *A. viscosus* preferentially deplete cysteine and asparagine, respectively, from culture media. Urea is present in relatively high concentrations (200 mg/litre) in saliva. Some oral species possess urease activity (e.g. *A. viscosus* and *S. salivarius*) and can convert urea to carbon dioxide and ammonia. At acidic pH, decarboxylation of amino acids would yield carbon dioxide and amines, while at high pH, deamination would produce ammonia and keto acids, which can be converted to acetic, propionic, and possibly iso- and n-butyric acids. For example, some periodontal pathogens can convert histidine, glutamine or arginine to acetate and butyrate. In this way, amino acid metabolism might be an important mechanism by which oral micro-organisms counter the extremes of pH caused by the catabolism of carbohydrates and urea.

Essential amino acids can be obtained from the environment or synthesized by the cell. Ammonia can be converted into a number of amino acids, for example:

$$\text{pyruvate} + NH_3 \underset{\displaystyle }{\overset{\displaystyle NADH + H^+ \quad NAD^+}{\rightleftharpoons}} \text{alanine} + H_2O$$

Further transamination reactions can provide other essential amino acids. Little is known at present about the mechanisms of transport of amino acids and peptides in oral bacteria. Peptides could be degraded into their constituent amino acids by intracellular proteases. One group of salivary peptides has been found to enhance the glycolytic activity of plaque bacteria. A peptide (termed sialin) with the structure glycine-glycine-lysine-arginine can give rise to a significant pH-rise effect in the laboratory and may, therefore, reduce the risk of caries development. Enhanced glycolysis leads to rapid clearance of sugars from the mouth, while base production both neutralizes the acids formed and raises the pH. A number of other salivary peptides can be metabolized by oral bacteria, and different species have different patterns of attack. S. sanguis strains are more proteolytic than mutans streptococci, although it has been claimed that saliva from subjects with a high incidence of caries is better at supporting the growth of S. mutans than saliva from caries-free people. A protein inhibitory to the growth of mutans streptococci has also been isolated from the saliva of some caries-free individuals. These proteins were anionic and did not appear to be related to other antimicrobial factors such as sIgA, IgG, lysozyme, lactoferrin, and sialo-peroxidase.

Many of the micro-organisms from the periodontal pocket are asaccharolytic but proteolytic, and depend for their growth on their ability to utilize the nutrients provided by GCF. During inflammation, many novel nutrients (e.g. haemoglobin, transferrin, haemopexin, haptoglobin, etc.) are provided by the host, and this can lead to the enrichment of highly proteolytic periodontopathogens such as P. gingivalis. These host molecules can be degraded to provide peptides and amino acids, as well as haeme, which is an essential co-factor for black-pigmented anaerobes. Different species appear to use particular amino acids for different purposes. For example, although both P. gingivalis and F.

nucleatum utilize glutamate, it was used in biosynthetic reactions by the former species and in catabolic pathways to generate energy by the latter. In addition to obtaining essential nutrients from GCF, these organisms are also able to degrade structural proteins and glycoproteins associated with the pocket epithelium. The production of enzymes such as chondroitin sulphatase, hyaluronidase, and collagenase, therefore, contributes to tissue damage and pocket formation. The pH optima of some of these enzymes are at neutral or slightly alkaline pH, which corresponds to that of the periodontal pocket. A small change in environmental pH can markedly alter the enzyme activity profile of some bacteria. In *P. gingivalis*, a rise in culture pH from 7.0 to 8.0 led to a change in the ratio of 'trypsin-like' to collagen breakdown activity from 1:1 to 8:1. These potential virulence factors are discussed in more detail in Chapter 7.

The metabolism of complex proteins leads to the generation of peptides of varying molecular sizes. Some of these released peptides can stimulate the growth of the proteinase-producing cell or of other neighbouring species. Some synergistic interactions appear to occur not only in the breakdown of these molecules (Chapter 5) but also at the molecular level. It has been shown that the low Eh necessary for the optimum activity of the main proteinase of *P. gingivalis* (gingivain; Chapter 7) could be met by the release of low molecular weight mercaptans from the catabolism of cysteine by *F. nucleatum*. Recent studies have also shown that some periodontopathogens such as *P. gingivalis* and *T. denticola* preferentially utilize peptides rather than amino acids. In these bacteria, peptide uptake is by a higher affinity uptake system than free amino acids. The optimum length of peptide for these transport systems was 10–14 residues; di- and tri-peptides were not taken up.

Collectively, these findings emphasize the significance of nitrogen metabolism in oral microbial ecology. Host and bacterial proteases are associated with tissue destruction in periodontal disease, while it has been argued that caries results not so much from an over-production of acid but more from a deficiency in base production by plaque bacteria.

Metabolism and inhibitors

Antimicrobial agents are used extensively in toothpastes and mouthrinses to help maintain dental plaque at levels compatible with oral health. Although they are often selected on the basis of a broad spectrum of antimicrobial activity, they frequently function in the mouth at sub-lethal concentrations. At such levels, these agents can interfere with carbohydrate and nitrogen metabolism.

Two of the most widely used inhibitors in dentistry can affect sugar transport by the PEP-PTS system of oral bacteria. While the primary caries-preventive action of fluoride is due to its effect on increasing the acid resistance of enamel (Chapter 6), fluoride at low concentrations is known to decrease the rate of sugar uptake and acid production by plaque bacteria. Fluoride has also been reported to reduce EPS pro-duction, and to inhibit the synthesis but not the breakdown of IPS (Figure 4.9). Fluoride inhibits enolase which converts 2-phosphoglycer-ate to phosphoenolpyruvate. This leads to glycolysis being inhibited directly and sugar transport being affected indirectly by a reduction in the availability of PEP for the PTS. Fluoride also inhibits ATP-synthase, and so interferes with the control of intracellular pH. Fluoride can also inhibit metallo-enzymes, e.g. phosphatases and phosphorylases, as well as oxidative enzymes such as catalase and peroxidase, and this may interfere with the redox balance within plaque.

The fluoride sensitivity of metabolism is pH-dependent, with the greatest inhibition occurring under acidic conditions. This is because at low pH fluoride exists as H^+F^- which is lipophilic and more easily able to penetrate membranes. The intracellular pH is relatively alkaline; therefore, once inside the cell, H^+F^- would dissociate and F^- would inhibit various enzymes as described above, while protons would acid-ify the cytoplasm and so tend to reduce:

1. the activity of glycolytic enzymes that have a pH optimum around neutrality,
2. the transmembrane pH gradient (and hence pmf-driven uptake and secretion processes),
3. the aciduricity of cells, for example, by interfering with the mem-brane-bound ATP synthase, which helps regulate the intracellular pH.

Chlorhexidine is widely used as a mouthwash to reduce plaque and prevent or treat gingivitis. It has a broad spectrum of antimicrobial activity but at sub-lethal levels it can affect many functions associated with the bacterial membrane. For example, it can abolish the activity of sugar transport by the PTS, and thereby severely reduce glycolysis. It can also inhibit the ATP synthase and affect the maintenance of ion gradients in streptococci. Chlorhexidine has also been found to interfere with nitrogen metabolism by inhibiting arginine uptake by *S. sanguis* and the trypsin-like enzyme of *P. gingivalis*.

A number of other antimicrobial agents have been shown to interfere with the metabolism of oral bacteria and, by so doing, play a major role in preventing disease. These agents, which include metal salts, plant extracts, and phenols, are discussed in detail in Chapters 6 and 7.

SUMMARY

Although the mouth is sterile at birth, the acquisition of the resident oral microflora begins within the first few hours of life. The biological properties of the mouth make it highly selective in terms of the types of micro-organisms able to colonize. Few of the species found in the mouths of adults and even fewer of the organisms of the general environment are able to establish successfully. Acquisition of the resident microflora follows a pattern of ecological succession: relatively few organisms (pioneer species) are able to colonize, but their presence enables other species to establish; this process eventually leads to a climax community with a high species diversity. Many species are acquired from the mother by transmission via saliva. The development of a climax community in the mouth can involve both allogenic (non-microbial influences) and autogenic (microbial influences) succession.

The composition of the resident microflora varies at different sites around the mouth, with each site having a relatively characteristic microbial community. Mutans streptococci and *S. sanguis* have preferences for hard surfaces for colonization, whereas species such as *S. salivarius* are recovered predominantly from the oral mucosa. The distribution of micro-organisms is related to their ability to adhere at a site, as well as to the need for their nutritional and environmental requirements (pH and redox potential) to be satisfied. Many species of bacteria have been shown to adhere by specific molecular interactions between adhesins located on their cell surface and ligands on the host; these ligands are derived mainly from the acquired pellicle and mucus coat on enamel and mucosal surfaces, respectively.

In order to cope with the fluctuating nutritional conditions in the mouth, the resident oral microflora is biochemically flexible. The primary source of nutrients is the endogenous supply of host proteins and glycoproteins from saliva and GCF. Superimposed on these are carbohydrates (and proteins) provided by the diet. Carbohydrates can be transported into the cell by, for example, a PEP-PTS system, and either converted to organic acids or used to synthesize IPS. Some disaccharides can be metabolized extracellularly into constituent sugars (for transport) or into EPS; these polysaccharides can be involved in attachment or used as extracellular storage compounds. The metabolism of nitrogen compounds involves the production of a wide range of exo- and endopeptidases; nitrogen metabolism can lead to base production which will help regulate environmental pH. The metabolism of oral micro-organisms is sensitive to many of the inhibitors used in preventive dentistry.

FURTHER READING

Cole, A. S. and Eastoe, J. E. (1988) *Biochemistry and Oral Biology*. Wright, London.

De Jong, M. H. and Van Der Hoeven, J. S. (1987) The growth of oral bacteria on saliva. *Journal of Dental Research*, **66**, 498–505.

Frandsen, E .V. G., Pedrazzoli, V. and Kilian, M. (1991) Ecology of viridans streptococci in the oral cavity and pharynx. *Oral Microbiology and Immunology*, **6**, 129–33.

Hamilton, I. R. (1987) Effects of changing environment on sugar transport and metabolism by oral bacteria, in *Sugar Transport and Metabolism in Gram-positive Bacteria*. (eds. A. Reizer and A. Peterkofsky) Ellis Horwood, Chichester, pp. 94–133.

Handley, P. S. (1990) Structure, composition and functions of surface structures on oral bacteria. *Biofouling*, **2**, 239–64.

Marsh, P. D. and Keevil, C. W. (1986) The metabolism of oral bacteria in health and disease, in *Microbial Metabolism in the Digestive Tract*. (ed. M. J. Hill) CRC Press, Boca Raton, pp. 155–81.

Percival, R. S., Challacombe, S. J. and Marsh, P. D. (1991) Age-related microbiological changes in the salivary and plaque microflora of healthy adults. *Journal of Medical Microbiology*, **35**, 5–11.

Theilade, E. (1990) Factors controlling the microflora of the healthy mouth, in *Human Microbial Ecology*. (eds. M. J. Hill and P. D. Marsh) CRC Press, Boca Raton, pp. 1–56.

Walker, G. J. and Jacques, N. A. (1987) Polysaccharides of oral streptococci, in *Sugar Transport and Metabolism in Gram-positive Bacteria*. (eds. A. Reizer and A. Peterkofsky) Ellis Horwood, Chichester, pp. 39–68.

5 Dental plaque

Dental plaque is a general term for the complex microbial community found on the tooth surface, embedded in a matrix of polymers of bacterial and salivary origin. Plaque that becomes calcified is referred to as calculus or tartar. The presence of plaque in the mouth can be demonstrated readily by rinsing with a disclosing solution such as erythrosin. The majority of plaque is found associated with the protected and stagnant regions of the tooth surface such as fissures, approximal regions and the gingival crevice (Figure 2.2).

DEVELOPMENT OF DENTAL PLAQUE

Cultural and microscopic techniques have been used to follow the process of bacterial succession on a tooth surface (Figure 5.1). The smooth surfaces of the anterior teeth have been chosen for most studies because of the accessibility of the sampling site. Bacteria rarely come into contact with clean enamel. As soon as a tooth surface is cleaned, salivary glycoproteins are adsorbed forming the acquired enamel pellicle (Chapter 4). Pellicle shows local variations in chemical composition and this can influence the pattern of microbial deposition. Large numbers of bacteria (up to 10^8 CFU ml^{-1}) are found in saliva. Unless swallowed, any of these organisms are likely to come into contact with a tooth surface. Indeed, colonization by many oral populations is related directly to their concentration in saliva.

Coccal bacteria are adsorbed onto the pellicle-coated enamel within two hours of cleaning. These pioneer species include *Neisseria* and streptococci, predominantly *S. sanguis*, *S. oralis*, and *S. mitis* (Table 5.1). *Actinomyces* spp. are also commonly isolated after two hours as are haemophili, but obligately anaerobic species are detected only rarely at this stage and usually in low numbers. These pioneer populations multiply, forming micro-colonies which become embedded in bacterial extracellular slimes and polysaccharides together with additional layers of adsorbed salivary proteins and glycoproteins (Figure 5.1 (a) and (b)).

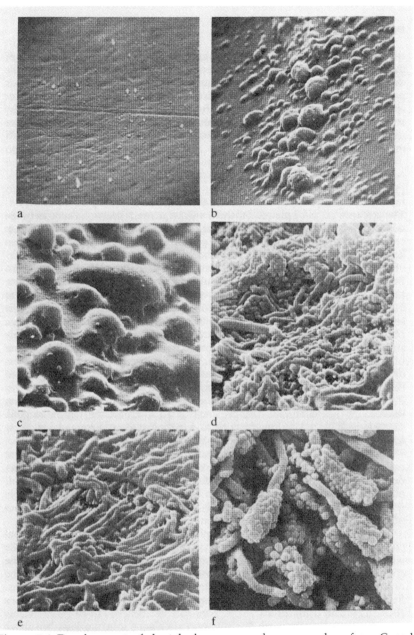

Figure 5.1 Development of dental plaque on a clean enamel surface. Coccal bacteria attach to the enamel pellicle as pioneer species (a) and multiply to form micro-colonies (b), eventually resulting in confluent growth (a biofilm) embedded in a matrix of extracellular polymers of bacterial and salivary origin (c). With time, the diversity of the microflora increases, and rod and filament-shaped bacteria colonize (d and e). In the climax community, many unusual associations between different bacterial populations can be seen, including 'corn-cob' formations (f). (Magnification approx. × 1150) Courtesy of C. A. Saxton.

Growth of individual micro-colonies eventually results in the development of a confluent film of micro-organisms (Figure 5.1 (c)–(e)). Growth
rates of bacteria are fastest during this early period with doubling times
ranging from 1–3 hours having been calculated. Studies with gnotobiotic
rats have shown that individual species can vary in their doubling times
(t_d); S. mutans and A. viscosus had t_d of 1.4 h and 2.7 h, respectively.
The proportions of the S. oralis-group increase during the first 48 hours
of plaque formation. The majority of isolates are S. oralis, S. mitis and
S. sanguis sensu stricto; fewer S. gordonii are recovered during these early
stages. As plaque develops into a biofilm so the metabolism of the
pioneer species creates conditions suitable for colonization by bacteria
with more demanding atmospheric requirements. Oxygen is consumed
by the aerobic and facultatively anaerobic species and replaced with
carbon dioxide. Gradually the Eh is lowered, which favours the growth
of obligately anaerobic species. Additional nutrients also become available and the diversity of the microflora increases both in terms of the
morphological types (Figure 5.1 (d)–(f)) and in the actual numbers of
species.

Table 5.1 Proportions of bacteria in developing supragingival plaque

	Time of plaque development (h)		
Bacterium	2	24	48
S. sanguis	8	12	29
S. oralis	20	21	12
mutans streptococci	3	2	4
S. salivarius	<1	<1	<1
S. viscosus	6	7	5
A. naeslundii	1	1	3
A. odontolyticus	2	3	6
Haemophilus spp.	11	18	21
Capnocytophaga spp.	<1	<1	<1
Fusobacterium spp.	<1	<1	<1
Black-pigmented anaerobes	0	<0.01	<0.1

The predominant bacteria colonizing the root surfaces have been
studied using an in situ appliance model system. Again the predominant
bacteria were streptococci (Table 5.2), particularly S. oralis, S. mitis, and
S. sanguis; S. salivarius was also a significant component of the microflora, although its proportions were found to decrease with time. Few

Gram-positive rods were recovered from the early stages, and *A. viscosus* was detected only after 24 hours (Table 5.2). Some species were not evenly distributed among the volunteers so that *A. viscosus* might predominate in one subject and *A. naeslundii* in another. Similarly, *Stomatococcus mucilagenosus* was found to be an early plaque colonizer in some people but not in others. Among other bacteria that were found occasionally were *Propionibacterium* spp. and *Rothia dentocariosa*. Gram-negative cocci (*Neisseria* and *Veillonella* spp.) were present but made up only a small fraction (approximately 2%) of the total cultivable micro-flora. Haemophili were also recovered but only occasionally and in low numbers. Figure 5.2 shows a palisading structure of dental plaque from a root surface.

Table 5.2 Proportions of bacteria colonizing pieces of tooth surface mounted in an intra-oral appliance

Bacterium	Time of plaque development (h)			
	4	8	12	24
S. sanguis	15	12	13	18
S. oralis	5	24	19	27
S. mitis	33	24	40	28
S. salivarius	19	12	7	2
A. naeslundii	3	1	1	2
A. viscosus	0	0	0	3

If plaque is allowed to accumulate undisturbed (as happens at stagnant sites) then there is a shift in the proportions of the bacteria within the biofilm. After 7 days, streptococci remain the dominant group of organisms but by 14 days, they constitute only around 15% of the cultivable microflora, and anaerobic rods and filaments predominate. The composition of the climax community varies depending on the site; the predominant bacteria will be described in a later section.

STRUCTURE AND FUNCTION OF DENTAL PLAQUE

The accumulation of plaque on teeth will be a result of the balance between deposition, growth and removal of micro-organisms. The development of plaque in terms of mass will continue until a critical size is reached. Shear forces will then limit any further expansion. However, structural development and re-organization may take place

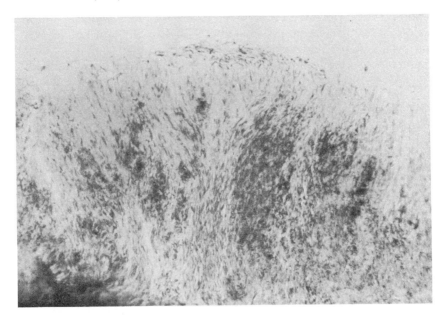

Figure 5.2 Structure of plaque growing on a root surface; evidence of palisading can be observed. (Magnification approx. × 400). Courtesy of K. M. Pang.

continually. Electron microscopy has demonstrated both a heterogeneous and a colonial type of sub-structure in sections of smooth surface plaque. The heterogeneous type is associated with palisaded regions where filaments and cocci appear to be aligned in parallel at right angles to the enamel surface. Micro-colonies, presumably of single populations, have also been observed. In addition, horizontal stratification has been described. The early stages of development results in a condensed layer of apparently a limited number of bacterial types. From 7 to 14 days, the bulk layer forms which shows less orientation but a higher morphological diversity. This layering has been attributed directly to bacterial succession. In mature plaque, organisms have been seen in direct contact with the enamel due to enzymic attack on the pellicle. Electron microscopy has confirmed the presence of an interbacterial matrix of polysaccharide. This matrix may function as a carbohydrate reserve and contribute towards the diffusion-barrier effect of plaque.

The structure and development of fissure plaque appears to be different to that on smooth surfaces. The microflora of fissures is less complex, being composed predominantly of rods and cocci. Palisading and

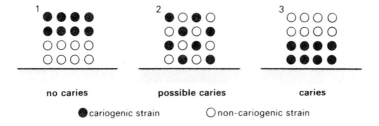

Figure 5.3 Diagram to show the importance of the spatial arrangement of an organism in dental plaque.

branching filaments are absent, although an inter-bacterial matrix can be observed. The pellicle is invariably degraded so that organisms are generally in direct contact with the enamel. Zones of impacted food particles are a common feature of fissures.

Three structurally-different types of plaque have been observed in the gingival crevice. Plaque from the gingival margin is similar to smooth surface plaque, while within the crevice two types, differing according to the size and structure of the condensed layer, have been recognized. Many bacterial associations can be observed in which cocci are arranged along the length of filamentous organisms. Such associations are described as 'corn-cob' and 'test tube brush' formations (Figure 5.1 (f)). The components of the 'corn-cob' will be described in a later section. The differences in plaque structure described above are supported by microbiological data, details of which will also be presented later.

The distribution of organisms within plaque has not been studied extensively. This is due to the technical difficulties of sample preparation and *in situ* isolate identification. Yet such spatial arrangements are of obvious importance, particularly in pathogenicity. A hypothetical situation to illustrate this point is shown in Figure 5.3. A caries-active strain known to comprise 50% of the cultivable microflora is shown in three possible spatial arrangements. It can be argued that example three with the caries-active strain directly over the enamel is potentially a more pathogenic situation than example one, where a layer of a caries-inactive organisms is acting as a barrier. Even this hypothetical model needs careful consideration since in example three potentially fermentable carbohydrate must be able to reach the caries-active strain. Indeed, bacteria with thickened cell walls have been observed on the enamel surface which might indicate unbalanced growth due to amino acid limitation. Difficulties in nutrient penetration might be due to the dif-

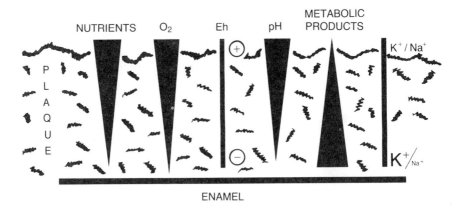

Figure 5.4 Schematic representation of the development of gradients in dental plaque.

fusion-limiting properties of plaque or to the metabolism of substrates nearer the surface by other organisms.

The diffusion-limiting property of plaque cannot be over emphasized. Either sharp or gentle gradients extending over small distances will exist in plaque for many of the parameters (physical and chemical) influencing microbial growth and survival. Sites close together may be vastly different in the types and concentrations of key nutrients, pH, Eh, and concentrations of toxic products of metabolism (Figure 5.4). Such vertical and horizontal stratifications will cause local environmental heterogeneity resulting in a collection of microhabitats or micro-environments. Each microhabitat potentially could support the growth of different populations and hence a different microbial community. Similarly, organisms residing in apparently the same general environment might be growing under quite dissimilar conditions. Thus when studying plaque, it is essential to take small samples from defined areas.

Summary

Plaque can be considered as a macrohabitat made up of many micro-environments due to gradients in ecologically significant parameters. The development and structure of plaque follows a definite pattern of bacterial colonization. Adhesion by pioneer species (*S. mitis*, *S. oralis*, *S. sanguis*) is followed by a gradual increase in the complexity of the microflora to a climax community of high species diversity, including many filamentous and obligately anaerobic bacteria. Variation and re-organization will occur within the micro-environment with time either due to host factors or as a result of the metabolism of the microbial community.

MECHANISMS OF PLAQUE FORMATION

The mechanisms of the formation of biofilms is currently attracting much interest in microbiology. This is because most micro-organisms in nature exist by attaching to a surface; these surface-associated microbes cause inestimable damage ranging from corrosion of pipes and fouling of ships to human infection. The accessibility of the mouth for study and the diversity of the plaque microflora has meant that the mechanisms by which oral organisms colonize the tooth surface have been investigated intensively for a number of years.

The development of a biofilm such as dental plaque can be divided arbitrarily into five stages. As a bacterium approaches a surface a number of specific and non-specific interactions will occur between the substratum and the cell which will determine whether attachment and colonization will take place. These interactions are:

1. van der Waals attractive forces, which operate over relatively long (>50 nm) separation distances.
2. At 10–20 nm distances, the interplay of van der Waals attractive forces and electrostatic repulsion produces a weak area of attraction that can result in reversible adhesion.
3. At these and shorter separation distances, the adhesion can become irreversible due to specific short range interactions between bacterial adhesins and host ligands.
4. The co-aggregation of bacteria to already attached cells.
5. The multiplication of the attached organisms to produce confluent growth and a biofilm.

These interactions will be described now in more detail but it should be remembered that biofilm formation is a dynamic process and that the five stages distinguished above are only arbitrary and are for the benefit of discussion. The attachment, growth, removal and reattachment of bacteria is a continuous and dynamic process and a microbial film such as plaque will undergo continuous reorganization.

For the specific adhesive interactions outlined in Chapter 4 to occur, the micro-organism and host surface must come into relatively close contact with each other. However, micro-organisms are negatively-charged due to the exposed molecules on their cell surface, while acidic proteins present in the acquired pellicle would also produce a net negative charge. The Derjaguin and Landau and the Verwey and Overbeek

(DLVO) theory has been used to describe the interaction between an inert particle (as a micro-organism might be envisaged at large separation distances) and a substratum. This theory states that the total interactive energy, V_T, of two smooth particles is determined solely by the sum of the van der Waals attractive energy (V_A) and the usually repulsive, electrostatic energy (V_R). Particles in aqueous suspension and surfaces in contact with aqueous solutions can acquire a charge due to, for example, the preferential adsorption of ions from solution or the ionization of certain groups attached to the particle or surface. The charge on a surface in solution is always exactly balanced by an equivalent number of counterions. Thus, the charge of a surface and the corresponding counterion charge in solution form an electrical double layer the size of which is inversely proportional to the ionic strength of the environment. As a particle approaches a surface, therefore, it experiences a weak van der Waals attraction induced by the fluctuating dipoles within the molecules of the two approaching surfaces. This attraction increases as the particle moves closer to the collector. However, as the surfaces approach each other a repulsive force is encountered due to the overlap of the electrical double layers. The magnitude of the repulsion will vary with the ionic strength and dielectric constant of the suspending medium and with the charge on the outer layers of the particle and on the surface to be colonized.

Curves can be plotted which show the variation of the total interactive energy, V_T, of a particle and a surface with the separation distance, h (Figure 5.5). A net attraction can occur at two values of h; these are referred to as the primary minimum (h very small) and the secondary minimum ($h = 10 - 20$ nm) and are separated by a repulsive maximum. The reversible nature of bacterial deposition suggests that the primary minimum is not usually encountered while the high ionic strength of saliva will increase the likelihood of oral bacteria experiencing a secondary minimum. Bacteria captured by a surface in this way are in equilibrium with the remaining organisms in the suspending medium and the number of captured cells will be dependent on the concentration of bacteria in suspension and on the depth of the secondary minimum.

The pattern of deposition could also be influenced by the type of interface from which the bacteria deposit. In many systems, bacteria adhere from a solid-liquid interface such as when a surface is immersed in a suspension of micro-organisms. In the mouth, however, a solid-liquid-air (s-l-a) interface is common which could be static but is more likely to be moving as a result of the continuous drainage of saliva over oral tissues. Small hydrophobic particles such as bacteria accumulate in the s-l-a contact zone. The concentration of bacteria in such a meniscus will be high and the surface area for deposition large improving the

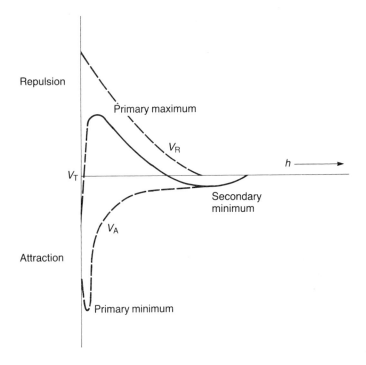

Figure 5.5. Diagram illustrating the DLVO theory. The total interactive energy, V_T, between a particle and a surface is shown with respect to the separation distance, h. The total interaction curve is obtained by the summation of an attraction curve, V_A, and a repulsion curve, V_R.

changes of bacterial attachment. Studies have shown that oral bacteria and polymeric material in saliva preferentially adsorb at the s-l-a interface and so the movement of these interfaces over the teeth may be important in the formation of dental plaque.

At separation distances of between 10–20 nm, therefore, organisms may be held irreversibly in a weak secondary minimum. With time, this interaction may become less reversible, or even irreversible due to adhesins on the cell surface becoming involved in specific short-range interactions. For this to occur, water films must be removed from between the interacting surfaces. It has been proposed that the major role of hydrophobicity and hydrophobic surface components is in their dehydrating effects on this water film enabling the surfaces to get closer together so that short range interactions can occur.

As discussed in Chapter 4, the short range forces involve interactions between specific components on the cell surface and ligands in the acquired pellicle. Examples include the interaction of bacterial lipotei-choic acid or GTF with blood group reactive substances (these have been characterized as sulphated glycoproteins) and between adsorbed antibodies (especially secretary IgA) and microbial antigens. A number of fractions of submandibular saliva have been shown to markedly affect the binding of a variety of oral bacteria to hydroxyapatite in laboratory studies. These fractions have been shown to include statherin and a high molecular weight acidic proline-rich protein (PRP-1). This PRP-1 fraction can promote the strong adhesion of *A. viscosus*, *A. israelii* and *A. odontolyticus*. Strains of *A. naeslundii* which only possess type 2 fimbriae (Chapter 4) do not respond to PRP-1. *S. mutans* (but not *S. sobrinus*), *Porphyromonas gingivalis*, *Prevotella loescheii* and *Prevotella melaninogenica* also display enhanced adsorption to PRP-1-treated hydroxyapatite whereas a strain of *Prevotella intermedia* and strains belonging to the *S. oralis*-group exhibited a variable response to PRP-1. Although adsorbed salivary components serve as receptors for *S. mutans*, the same com-ponents do not facilitate the attachment of *S. sobrinus*. The adsorbed glucan/GTF fraction appears to promote the attachment of *S. sobrinus* to hydroxyapatite; this mechanism may also enhance the attachment of other oral bacteria. A significant finding was that although *A. viscosus* could bind avidly to PRP-1 adsorbed on to a surface, it did not interact with this protein in solution. It has been proposed that hidden molecu-lar segments of PRPs become exposed, as a result of conformational changes, when the protein was adsorbed to hydroxyapatite. Such hidden receptors for bacterial adhesins have been termed 'cryptitopes'.

Adhesins which recognize cryptitopes in surface-associated molecules would provide a strong selective advantage for any micro-organism which colonizes a mucosal or tooth surface. The secretions bathing these surfaces contain components which are structurally related to those on the surfaces of mucosal cells. This molecular mimickry is thought to contribute to the cleansing actions of such secretions. When in solution, these components can bind to organisms and cause aggregation; larger aggregates are more easily lost from the mouth by swallowing and this facilitates bacterial clearance. However, if these components become altered in configuration upon adsorption to a sur-face, then the newly-exposed cryptitopes would promote the coloniz-ation of the surface by specific micro-organisms. Another example of a cryptitope involving conformational change is the binding of members of the *S. oralis*-group to fibronectin when complexed to collagen but not to fibronectin in solution. It has been suggested that this might be a mechanism whereby certain oral streptococci are able to colonize dam-aged heart valves in endocarditis.

Another example of a cryptitope involved in plaque formation is the recognition of galactosyl-binding lectins by oral bacteria. Epithelial cells and the acquired enamel pellicle have mucins with oligosaccharide side chains with a terminal sialic acid. Bacteria such as *A. viscosus* and *A. naeslundii* synthesize neuraminidase which cleaves the sialic acid exposing the penultimate galactosyl sugar residue. Many oral bacteria possess galactosyl-binding lectins including *A. viscosus, A. naeslundii, L. buccalis, F. nucleatum, E. corrodens* and *P. intermedia*, and would benefit from the exposure of these cryptitopes. Similarly, the binding of *P. gingivalis* is greater to epithelial cells that have been mildly treated with trypsin. Periodontopathogens including *P. gingivalis* produce 'trypsin-like' proteases and it has been proposed that the metabolism of these bacteria may create appropriate cryptitopes for their colonization. Following on from this, it has also been suggested that poor oral hygiene can generate new cryptitopes by encouraging plaque accumulation which can lead to the emergence of highly proteolytic bacteria.

Once the pioneer bacteria have attached, then growth and micro-colony formation will occur. The early colonizers include members of the *S. oralis*-group; the production of an IgA$_1$ protease by these pioneer species may be significant in their ability to survive and proliferate during the early stages of plaque formation. *S. oralis* and some strains of *S. mitis* can produce neuraminidase and IgA$_1$ protease and these species can comprise up to 27% and 42% of the plaque microflora, respectively, after 24 hours. Salivary polymers will continue to be adsorbed on to bacteria already attached to the tooth surface (Figure 5.6) and therefore several of the mechanisms described above will continue to operate.

An important mechanism that aids plaque build-up and encourages species diversity is the phenomenon of coaggregation between microbial cells. Bacterial accumulation will be accelerated by intrageneric coaggregation among streptococci and among actinomyces, as well as by intergeneric coaggregation between streptococci and actinomyces. The subsequent development of dental plaque will involve intergeneric coaggregation between other genera and the primary colonizers. For example, coaggregation can occur between:

Gram-positive species

e.g.   *S. sanguis* or   and *Actinomyces* spp.,
      S. mitis   *Corynebacterium matruchotii*, or
                              Propionibacterium acnes

Gram-negative species

e.g.   *Prevotella melaninogenica* and *Fusobacterium nucleatum*

Gram-positive and Gram-negative species

e.g.   *Streptococcus* spp. or   and   *Prevotella* spp.,
       Actinomyces spp.   *Capnocytophaga* spp.,
                                             Fusobacterium nucleatum,
                                             Eikenella corrodens,
                                             Veillonella spp., or
                                             Porphyromonas gingivalis

Coaggregation is a common feature of oral streptococci and actino-myces; of 300 strains surveyed, 84% were able to coaggregate. Based on the nature of the interaction, six coaggregation groups of streptococci (1–6) and actinomyces (A–F) have been described. Intrageneric coaggre-gation appears to be limited mainly to early plaque colonizers, especially

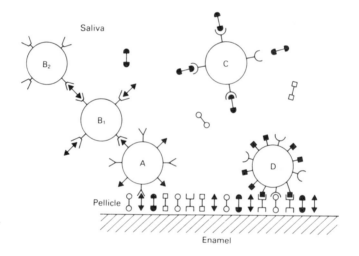

Figure 5.6 A diagrammatic representation of the attachment of bacteria from saliva to pellicle-coated enamel. The pattern of attachment is influenced by the composition of the pellicle (ligands) and the surface components of the bacteria (adhesins). The diagram illustrates the role of specific molecular interactions both between enamel pellicle and cells A and D, and in the co-aggregation of cell B_1 to cell A. Salivary components can both promote plaque formation by, for example, facilitating adhesion between cells B_2 and B_1, and also prevent plaque attachment by saturating receptors on the bacterial surface, as with cell C. Salivary components can also prevent adhesion by causing aggregation; large aggregates of cells are more easily lost from the mouth by swallowing (see the text).

streptococci. In contrast, fusobacteria, which coaggregate with the widest range of genera, do not coaggregate with other fusobacteria.

Coaggregation involves lectins; these carbohydrate-binding proteins interact with the complementary carbohydrate-containing receptor on another cell. Thus, the lectin-mediated interaction between streptococci and actinomyces can be blocked by adding galactose or lactose or by treating the receptor with a protease. The process of coaggregation can also lead to the development of the 'corn-cob' structures (Figure 5.1 (f)). 'Corn-cobs' can be formed between 'tufted-fibril' streptococci (Chapters 3 and 4) and *C. matruchotii*; similar associations have also been found between *Eubacterium* and *Veillonella* spp.

Detailed studies of the coaggregation of *A. viscosus/naeslundii* with *S. sanguis* and *S. mitis* have identified three types of interaction. One involves a lectin on the fimbriae of the Gram-positive rod and a receptor on the streptococci; the interaction is inhibited by galactose or lactose and by heat or protease treatment of the *Actinomyces* but not of the streptococci. The second type of interaction involves a heat- or protease-sensitive lectin on the streptococci binding to 'treatment-resistant' receptors on the *Actinomyces*, while the third involves labile components on both cell types. It has been postulated that lectin-mediated coaggregation is an important mechanism in the structural organization of microbial communities such as dental plaque. The development of food-chains including those between streptococci and veillonellae and the formation of the 'corn-cob' structures might be facilitated by lectin-like interactions. Many of the coaggregation reactions described above have been determined in the laboratory in the absence of saliva. Whether salivary molecules would enhance or block some of these interactions has yet to be established.

Another factor in the development of plaque will be the synthesis of extracellular polymers by adherent bacteria. In particular, the synthesis of water-insoluble mutan by *S. mutans* will make an important contribution to the structural integrity and diffusion-limiting properties of plaque. Originally, it was believed that polysaccharide production was important in the initial stages of attachment. Studies with mutants demonstrated strains of *S. mutans* unable to produce water-insoluble glucans from sucrose produced less plaque and fewer caries lesions on smooth enamel surfaces than did the parent strains. Although preformed polymer could be involved in colonization, the role of polysaccharide production is now regarded to be more important in consolidating the attachment of cells already on a surface and in the development of plaque as a biofilm.

The relative importance of individual mechanisms of bacterial attachment in the mouth is obviously a contentious issue and one that is difficult to resolve. For example, while some believe that wall and/or membrane-associated teichoic acid interactions with pellicle components are of major importance, others point to the observations that many teichoic acid-producing bacteria do not adhere while teichoic acid-deficient organisms can attach quite readily. The diversity of potential mechanisms for adherence together with the molecular heterogeneity of the microbial and host surface probably means that binding involves multiple interactions.

Plaque development also involves the concurrent growth of the attached bacteria to produce a microbial film. This aspect of plaque development has received little attention to date but a consideration of the literature on surface-associated growth of micro-organisms from other ecosystems suggests that the physiology of bacteria in plaque might be different to that predicted from studies of the organisms in liquid cultures. The sensitivity of surface-associated micro-organisms (particularly when in a biofilm) to a range of antibiotics and disinfectants is reduced compared to the same organisms in solution. It is yet to be determined if the sensitivity of oral organisms to the antimicrobial agents being incorporated into new mouthwashes and toothpastes will be affected when they are growing in dental plaque. Early studies of *Bacillus* spp., enteric and marine micro-organisms in nutrient-depleted environments showed that their metabolic activity could be raised in the presence of a surface. Lactic acid production by oral streptococci has been found to be enhanced by the addition of hydroxyapatite crystals to their growth medium while continuous culture studies of a pseudomonad from an estuarine source showed that when microscope slides were immersed in the liquid culture the growth rate of the organisms that adhered to the surface was always greater than that imposed on the cells by the chemostat. The organism grew on the surface as discrete micro-colonies before eventually developing into a film. When similar experiments were performed on mixed cultures from either river water or dental plaque, the communities recovered from the surface were always different, both in the types and numbers of species present, from those in the respective bulk fluid. Taken together, these results suggest that a surface provides a different environment for growth, and that surface-associated rates of growth can be faster than those obtained in liquid culture. Several hypotheses have been proposed to explain the initial enhanced growth of bacteria on a surface. One widely accepted theory is that there is an increased concentration of nutrients at an interface between a surface and a liquid. Another explanation is based on the hypothesis that energy-conserving and energy-sharing interactions between bacterial cells would be more likely to occur on a surface, particularly in a microbial film. Studies of ecosystems as diverse as the gastrointestinal tract and marine sediments have recovered bacteria able to utilize end products of metabolism such as lactate in close association with lactate-producers. The molecular understanding of surface-enhanced bacterial growth remains an important but contentious area and more research will be necessary to provide data to support the hypotheses outlined above.

Summary

The adhesion of organisms to the tooth surface is a complex process involving, initially, weak electrostatic attractive forces, followed by a variety of polymer bridging interactions; these processes together with the synthesis of extracellular polysaccharides from, for example, sucrose serve to increase the probability of permanent attachment. Pioneer species interact directly with the acquired pellicle while biofilm formation is dependent on intra- and intergeneric coaggregation between bacteria (involving lectin-mediated binding) and the growth of the attached micro-organisms.

BACTERIAL COMPOSITION OF THE CLIMAX COMMUNITY OF DENTAL PLAQUE FROM DIFFERENT SITES

Environmental conditions on a tooth are not uniform. Differences exist in the degree of protection from oral removal forces and in the gradients of many biological and chemical factors that influence the growth of the resident microflora. These differences will be reflected in the composition of the microbial community, particularly at sites so obviously distinct as the gingival crevice, approximal regions, smooth surfaces, and pits and fissures.

Fissure plaque

The microbiology of fissure plaque has been determined using either 'artificial fissures' implanted in occlusal surfaces of pre-existing restorations, or by sampling 'natural' fissures. The microflora is mainly Gram-positive and is dominated by streptococci, especially extracellular-polysaccharide producing species. In one study no obligately-anaerobic Gram-negative rods were found, while others have recovered anaerobes including *Veillonella* and *Propionibacterium* species in low numbers (Table 5.3). *Neisseria* spp. and *Haemophilus parainfluenzae* have also been isolated on occasions. A striking feature of the microflora is the wide range of numbers and types of bacteria in the different fissures. In one study, the total anaerobic microflora ranged from 1×10^6 to 33×10^6 CFU per fissure. This suggests that the ecology of each fissure might be different. The factors that determine the final composition of the microflora in fissures are not known, but the influence of saliva at this site must be of great significance. The simpler community found in fissures compared to other enamel surfaces probably reflects a more severe environment, perhaps with a more limited range of nutrients. The distribution of bacteria within a fissure has not been studied in detail, although it has been claimed that lactobacilli and mutans streptococci preferentially inhabit the lower depths of a fissure. It is clear from Figure 5.7 that environmental conditions at the base of the fissure will be very different in terms of nutrient availability, pH, buffering effects of saliva, etc., than other areas nearer the plaque surface.

Table 5.3 The predominant cultivable microflora of occlusal fissures in ten adults

Bacterium	Median percentage of total cultivable microflora	Range	Percentage isolation frequency
Streptococcus	45	8–86	100
Staphylococcus	9	0–23	80
Actinomyces	18	0–46	80
'Arachnia'*	2	0–21	60
Propionibacterium	1	0–8	50
Eubacterium	0	0–27	10
Lactobacillus	0	0–29	20
Veillonella	3	0–44	60
Individual species			
mutans streptococci	25	0–86	70
S. sanguis	1	0–15	50
S. oralis	0	0–13	30
'S. milleri'-group	0	0–3	10
A. naeslundii	3	0–44	70
A. viscosus	3	0–17	80
L. casei	0	0–10	10
L. plantarum	0	0–29	10

* *Arachnia propionica* has now been reclassified as *Propionibacterium propionicus*.

Figure 5.7 Dental plaque in a fissure on the occlusal surface of a molar. (Magnification approx. × 100). Courtesy of K. M. Pang.

Approximal plaque

The main organisms isolated from a study of approximal plaque are shown in Table 5.4. Although streptococci are present in high numbers, these sites are frequently dominated by Gram-positive rods, particularly *Actinomyces* spp. The more reduced nature of this site compared to that of fissures can be gauged from the higher recovery of obligately anaerobic organisms although spirochaetes are not commonly found. Again, the range of counts of most bacteria is high, suggesting that each site represents a distinct ecosystem which should be looked at in isolation with regard to the relationship between the resident microflora and the clinical state of the enamel.

Table 5.4 The predominant cultivable microflora of approximal plaque

Bacterium	Mean percentage of total cultivable microflora	Range	Percentage isolation frequency
Streptococcus	23	0.4–70	100
Gram-positive rods (predominantly *Actinomyces*)	42	4–81	100
Gram-negative rods (predominantly 'Bacteroides')	8	0–66	93
Neisseria	2	0–44	76
Veillonella	13	0–59	93
Fusobacterium	0.4	0–5	55
Lactobacillus	0.5	0–2	24
Rothia	0.4	0–6	36
Individual species			
mutans streptococci	2	0–23	66
S. sanguis	6	0–64	86
S. salivarius	1	0–7	54
'S. milleri'-group	0.5	0–33	45
A. israelii	17	0–78	72
A. viscosus/naeslundii	19	0–74	97

Viable counts were derived from 58 samples of approximal plaque from 10 schoolchildren.

The variability of plaque has again been highlighted in a recent study in which several small samples have been taken from different sites around the contact area of teeth extracted for orthodontic purposes. An example is shown in Table 5.5; at each approximal site the total numbers, as well as the range and types, of bacteria vary, again emphasizing the need for accurate sampling of discrete sites when attempting to correlate the composition of plaque with disease.

Table 5.5 The cultivable microflora from three sites on the approximal surface of an extracted tooth from a schoolchild

Bacterium	Viable count (colony forming units)		
	site 1	site 2	site 3
Total count	1.7×10^4	6.8×10^2	7.9×10^5
Streptococcus	0	0	6.1×10^5
Actinomyces	5.8×10^2	0	1.8×10^5
Neisseria	25	25	1.5×10^4
Veillonella	0	0	6.5×10^4
Capnocytophaga	0	0	1.3×10^2
Haemophilus	1.7×10^4	6.0×10^2	0
Individual species			
mutans streptococci	0	0	3.7×10^4
S. sanguis	0	0	1.1×10^5
S. oralis	0	0	1.0×10^4
S. salivarius	0	0	1.8×10^4
A. viscosus	5.8×10^2	0	1.8×10^5
A. naeslundii	0	0	6.5×10^3
Gram-negative 'spreading' filament	0	+	0

Gingival crevice plaque

An obviously distinct ecological climate is found in the gingival crevice. This is reflected in the higher species diversity of the bacterial community at this site although the total numbers of bacteria can be low (10^3–10^6 CFU/crevice). In contrast to the microflora of fissures and approximal surfaces, higher levels of obligately anaerobic bacteria can be found, many of which are Gram-negative (Table 5.6). Indeed, spiro-

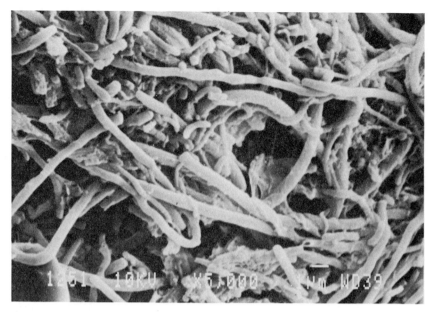

Figure 5.8 Scanning electron micrograph of subgingival plaque, showing rods, curved rods, filaments and spiral-shaped cells. (Magnification approx. × 5000). Courtesy of K. M. Pang.

chaetes and anaerobic streptococci are isolated almost exclusively from this site (Figure 5.8). The ecology of the crevice is influenced by the anatomy of the site and the flow and properties of GCF. Many organisms that are asaccharolytic but proteolytic are found in the gingival crevice; they derive their energy from the hydrolysis of host proteins and peptides and from the catabolism of amino acids. In disease the gingival crevice enlarges to become a periodontal pocket (Figure 2.1) and the flow of GCF increases. The diversity of the microflora increases still further and will be described in more detail in Chapter 7. Among the genera and species associated with the healthy gingival crevice are members of the *S. oralis* and '*S. milleri*'-group (especially *S. anginosus*, Table 4.4); in addition, *Actinomyces meyeri, A. odontolyticus, A. viscosus, A. naeslundii, A. georgiae, A. gerencseriae Capnocytophaga ochracea* and *Rothia dentocariosa* can also be found. The most commonly isolated black-pigmented anaerobe in the healthy gingival crevice is *P. melaninogenica* while *P. intermedia* has also been recovered on occasions; *P. gingivalis* is rarely isolated from healthy sites. Fusobacteria can also be found in the healthy gingival crevice.

Table 5.6 The predominant cultivable microflora of the healthy gingival crevice

Bacterium	Mean percentage of total cultivable microflora	Range	Percentage isolation frequency
Gram-positive facultatively anaerobic cocci (predominantly *Streptococcus*)	40	2–73	100
Gram-positive obligately anaerobic cocci (predominantly *Peptostreptococcus*)	1	0–6	14
Gram-positive facultatively anaerobic rods (predominantly *Actinomyces*)	35	10–63	100
Gram-positive obligately anaerobic rods	10	0–37	86
Gram-negative facultatively anaerobic cocci (predominantly *Neisseria*)	0.3	0–2	14
Gram-negative obligately anaerobic cocci (predominantly *Veillonella*)	2	0–5	57
Gram-negative facultatively anaerobic rods	ND	ND	ND
Gram-negative obligately anaerobic rods	13	8–20	100

Samples were taken from the gingival crevice of seven adult humans
ND, not detected.

Denture plaque

The microflora of denture plaque from healthy sites (i.e. with no sign of denture stomatitis; Chapter 8) is highly variable as can be deduced from the wide ranges in viable counts obtained for individual bacteria shown in Table 5.7. Clear differences are also apparent between the fitting and the exposed surfaces of the denture. In the relatively stagnant area on the denture-fitting surface, plaque tends to be more acidogenic, thereby favouring streptococci (especially mutans streptococci) and sometimes *Candida* spp. In edentulous subjects, dentures become the primary habitat for mutans streptococci and *S. sanguis*. It has been claimed that the microflora of denture plaque overlying healthy palatal mucosa is similar to that of fissure plaque in that streptococci, actinomyces and sometimes lactobacilli predominate. However, denture plaque can contain large proportions of the anaerobe *A. israelii*, and is also characterized by having extremely low levels of Gram-negative rods. Interestingly, *S. aureus* was regularly isolated (Table 5.7) in one study of denture plaque. This species is also found commonly on the mucosa of patients with denture stomatitis (Chapter 8).

Dental plaque from animals

There is interest in the microbial composition of dental plaque from animals for two main reasons: (a) to study the influence of widely different diets and life-styles on the microflora, and (b) to determine the similarity between the microflora of an animal with that of humans to ascertain their relevance as a model of human oral diseases. At the genus level, the plaque microflora is similar among animals representing such diverse dietary groups as insectivores, herbivores and carnivores. This again emphasizes (a) the significance of endogenous nutrients in maintaining the stability and diversity of the resident microflora and (b) the specificity of the interaction between this microflora and the oral ecosystem. Most studies have been of animals kept in captivity; there have been few investigations into the composition of the oral microflora of animals in the wild. *Actinomyces, Streptococcus, Neisseria, Veillonella,* and *Fusobacterium* are widely distributed in both zoo and non-zoo primates and other animals, and can, therefore, be genuinely considered as autochthonous members of dental plaque (Chapter 4). Following recent taxonomic studies, differences between isolates from man and animals have emerged at the species level. For example, *S. rattus* and *S. macacae* are isolated exclusively from rodents and primates, respectively, whereas other mutans streptococcal species

Table 5.7 The predominant cultivable microflora of denture plaque

Micro-organism	Percentage viable count		Percentage isolation frequency
	Median	Range	
Streptococcus	41	0–81	88
mutans streptococci	< 1	0–48	50
S. sanguis	1	0–4	63
S. oralis	2	0–30	75
'S. milleri'-group	2	0–51	63
S. salivarius	0	0–41	38
Staphylococcus	8	1–13	100
S. aureus	6	0–13	88
'S. epidermidis'	0	0–7	13
Gram-positive rods	33	1–74	100
Actinomyces	21	0–54	88
A. israelii	3	0–47	63
A. viscosus	<1	0–48	50
A. naeslundii	3	0–11	63
A. odontolyticus	1	0–17	63
Lactobacillus	0	0–48	25
Propionibacterium	<1	0–5	50
Veillonella	8	3–20	100
Gram-negative rods	0	0–6	38
Yeasts	0.002	0–0.5	63

are found in humans (Table 3.3). Other species differences between man and animals were also highlighted in Chapter 3.

PLAQUE FLUID

Plaque fluid is the free aqueous phase of plaque, and can be separated from the microbial components by centrifugation. Analysis of plaque fluid shows it to differ in composition from both saliva and GCF. In particular, the protein content of plaque fluid is higher than that of saliva, as are the concentrations of several important cations including sodium, potassium, and magnesium. Likewise, the levels of albumin, lactoferrin, and lysozyme are greater in plaque fluid than saliva, although this trend is reversed for amylase. A number of enzymes of both bacterial and host (e.g. from polymorphs) origin can be detected in plaque fluid. Specific host defence factors are also found in plaque fluid; sIgA is present at the same concentration as in whole saliva whereas IgG and complement are at higher levels, and are probably derived from GCF. Fluoride binds to plaque components, but is also found free in plaque fluid. Bound fluoride can be released from these components because fluoride concentrations increase in plaque fluid when the pH falls during the bacterial metabolism of fermentable carbo-hydrates. Acidic products of metabolism are retained in plaque fluid, and shifts in their profile from a hetero- to a homofermentative pattern can be observed following the intake of dietary sugars. The ratio of K^+/Na^+ is higher in plaque fluid, and this may have a significant influence on several properties of oral streptococci associated with their role in disease. As discussed in Chapter 4, these properties include enhanced acid production and the increased secretion of glycosyltransferases by potasssium ions, possibly by the stimulation of the protonmotive force in cells.

Analysis of plaque fluid has allowed insights into nitrogen meta-bolism by oral bacteria. Complex patterns of amino acid consumption and excretion occur over relatively short time periods. Hydroxyproline and hydroxylysine have been found consistently in human plaque fluid which may indicate collagen breakdown as being a significant activity in plaque. Likewise, a novel nitrogenous compound, delta-aminovaleric acid, has been found in plaque fluid from humans and monkeys, and is probably derived from proline by reductive ring cleavage and deamin-ation. These amino acids and proteins, together with the anions and organic acids present, will contribute to the buffering capacity of plaque fluid, which is twice that of resting saliva.

CALCULUS

Calculus, or tartar, is the term used to describe calcified dental plaque. It consists of intra- and extracellular deposits of mineral, including apatite, brushite, and whitlockite, as well as protein and carbohydrate. Mineral growth can occur around any bacteria; areas of mineral growth can then coalesce to form calculus which may become covered by an unmineralized layer of bacteria. Calculus can occur both supragingivally (especially near the salivary ducts) and subgingivally, where it may act as an additional retentive area for plaque accumulation, thereby increasing the likelihood of gingivitis and other forms of periodontal disease. Calculus can be porous leading to the retention of bacterial antigens and the stimulation of bone resorption by toxins from periodontopathogens. Over 80% of adults have calculus, although its prevalence increases with age. An elevated calcium ion concentration in saliva may predispose some individuals to be high calculus formers. Once formed, huge removal forces are required to detach calculus; this removal takes up a disproportionate amount of clinical time during routine visits by patients to the dentist. Consequently, a number of dental products are now formulated to restrict calculus formation. These products contain pyrophosphates, zinc salts, or polyphosphonates to inhibit mineralization by slowing crystal growth and reducing coalescence.

MICROBIAL INTERACTIONS IN DENTAL PLAQUE

In a biofilm such as dental plaque, micro-organisms are in close proximity to one another and interact as a consequence. These interactions can be beneficial to one or more of the interacting populations, while others can be antagonistic (Table 5.8). Microbial metabolism within plaque will produce gradients in factors affecting the growth of other species. These would include the depletion of essential nutrients with the simultaneous accumulation of toxic or inhibitory by-products (Figure 5.4). As stated earlier, these gradients lead to the development of vertical and horizontal stratifications within the plaque biofilm. Such environmental heterogeneity has two important consequences for microbial interactions. It enables organisms with widely differing requirements to grow, and ensures the co-existence of species that would be incompatible with one another in a homogeneous habitat. Most studies have characterized microbial interactions in the laboratory, with the assumption that they will operate similarly in the mouth. Particularly useful studies have involved laboratory model systems, such as the artificial mouth and the chemostat, and gnotobiotic or specific pathogen-free animals.

Table 5.8 Factors involved in microbial interactions in dental plaque

Beneficial	Antagonistic
Enzyme complementation	Bacteriocins
Food chains (food webs)	Hydrogen peroxide
Coaggregation	Organic acids
	Low pH
	Nutrient competition

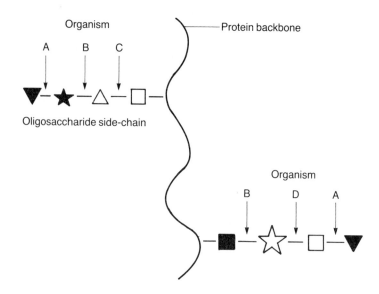

Figure 5.9 Bacterial co-operation in the degradation of host glycoproteins (enzyme complementation). For example, organism A is able to cleave the terminal sugar of the oligosaccharide side-chain, which enables organism B or D to cleave the penultimate residue, etc.

Competition for nutrients will be one of the primary ecological determinants in dictating the prevalence of a particular species in dental plaque. Bacterial growth in most natural ecosystems is limited by the availability of essential nutrients and co-factors, and there is now considerable evidence that the growth of the resident oral microflora is also dependent on the utilization of endogenous nutrients. The plaque microflora possesses a wide range of proteases and glycosidases, suggesting that salivary proteins and glycoproteins are the major sources of nitrogen and carbon at healthy sites. Individual species of oral bacteria possess different but overlapping patterns of enzyme activity, so that the concerted action of several species is necessary for the complete degradation of host molecules. This is illustrated diagrammatically in Figure 5.9; it can be seen that the growth of some organisms will be dependent on others removing, for example, the terminal sugar from the oligosaccharide side-chain of the glycoprotein.

An example of microbial co-operation in the breakdown of host macromolecules has been provided by a study of the enrichment of subgingival bacteria on human serum (used to mimic GCF). Prolonged growth of subgingival plaque in a chemostat led to the selection of consortia of bacteria with different metabolic capabilities. Shifts in the microbial composition of the consortia occurred at different stages of glycoprotein breakdown. Initially, carbohydrate side-chains were removed by organisms with complementary glycosidase activities, including *S. oralis*, *E. saburreum* and *Prevotella* spp. This was followed by the hydrolysis of the protein core by, for example, *P. intermedia*, *P. oralis*, *F. nucleatum*, and to a lesser extent, *Eubacterium* spp.; some amino acid fermentation occurred and the remaining carbohydrate side-chains were metabolized leading to the emergence of *Veillonella* spp. A final phase was characterized by progressive protein degradation and extensive amino acid fermentation; the predominant species included *Peptostreptococcus micros* and *E. brachy*. Significantly, individual species grew only poorly in pure culture in serum. A consequence of these interactions involving enzyme complementation is that different species avoid direct competition for individual nutrients, and hence are able to co-exist. This type of interaction is an example of **proto-cooperation** or **mutualism**, whereby there is benefit to all participants that are involved in the interaction. Bacterial polymers are also targets for degradation. EPS synthesized by many plaque bacteria (Table 4.8) can be metabolized by other organisms in the absence of exogenous (dietary) carbohydrates. The fructan of *S. salivarius* and other streptococci, and the glycogen-like polymer of *Neisseria*, are particularly labile, and little fructan can be detected in plaque *in vivo*. In addition, mutans streptococci, members of the *S. oralis*-group, *S. salivarius*, *A. israelii*, *Capnocytophaga* spp., and *Fusobacterium* spp. possess exo- and/or endohydrolytic activity and metabolize streptococcal glucans. The metabolism of these polymers is usually cited as an example of **commensalism** (an interaction beneficial to one organism but with a neutral effect on the other). As with the degradation of salivary proteins and glycoproteins, proto-cooperation may occur if organisms with exo-dextranase activity combine with an endo-dextranase producer to attack polymers to their mutual benefit. Some have reported that polymer metabolism might lead to **antagonism** (inhibition of one or more of the interacting species). For example, dextranase-producing *S. oralis* strains were able to block the sucrose-mediated aggregation of mutans streptococci and *S. sanguis* to each other, and the adherence of each species to a solid surface. This antagonistic interaction was accomplished through competition for cell surface glucan binding sites, the destruction of glucans capable of functioning as cell binding sites, and the suppression of water-soluble glucan synthesis from sucrose.

Many other types of nutritional interactions are known to exist among plaque micro-organisms whereby the products of metabolism of one organism (primary feeder) is the main source of nutrients for another (secondary feeder). The best described interaction of this type is the utilization of lactate produced from the metabolism of dietary carbohydrates by a range of other species, but particularly *Streptococcus* and *Actinomyces* species. This food chain was demonstrated originally in mixed continuous culture studies in which *S. mutans* acted as the primary feeder. This interaction could have important ecological consequences. Lactic acid is the strongest acid produced in quantity in dental plaque and is, in consequence, implicated in enamel demineralization. By converting lactate into propionic and acetic acids, which are weaker, it has been proposed that *Veillonella* spp. could reduce the cariogenic potential of other plaque bacteria. Evidence to support this concept came from gnotobiotic animal studies. Fewer caries lesions were obtained in rats inoculated with either *S. mutans* or *S. sanguis* and *Veillonella* than in animals infected with either streptococcus alone:

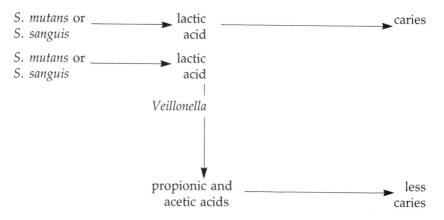

Strains of *Neisseria*, *Corynebacterium*, and *Eubacterium* are also able to metabolize lactate. However, some doubts on the clinical relevance of this interaction between mutans streptococci and *Veillonella* have been expressed recently. Laboratory studies have now shown that this interaction is an example of proto-cooperation because the continual consumption of the end product of metabolism actually enhances the growth and the rate of glycolysis and acid production of the lactate-producing strain. Similarly, in an artificial mouth model, demineralization of enamel pieces by *S. mutans* was increased in the presence of *Veillonella*. Furthermore, some recent human clinical studies have reported increases in both mutans streptococci and *Veillonella* at sites with caries (Chapter 6).

Other nutritional interactions between oral bacteria have been described in the laboratory. Under anaerobic conditions, a strain of *S.*

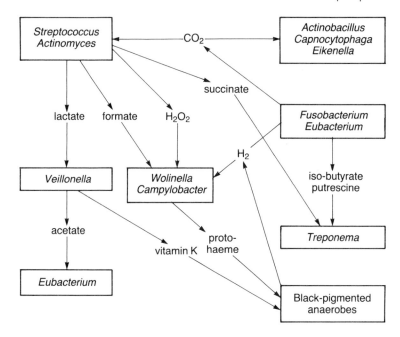

Figure 5.10 Some potential nutritional interactions (food chains) between plaque bacteria.

mutans required *p*-aminobenzoic acid for growth, and this could be supplied by *S. sanguis*. Oral spirochaetes can be dependent on the production of iso-butyrate and putrescine or spermine by fusobacteria, and on succinate produced by various Gram-positive rods. Similarly, *Fusobacterium* and *Prevotella* species provide hydrogen and formate for the growth of *Wolinella* and *Campylobacter* spp., while the metabolism of some black-pigmented anaerobes is dependent on the synthesis of vitamin K by other bacteria in the gingival crevice. A mutually beneficial interaction between *S. sanguis* and *C. rectus* has been described whereby the anaerobe scavenges inhibitory oxygen, or possibly hydrogen peroxide produced by the streptococcus, while *S. sanguis* provides *C. rectus* with formate following the fermentation of glucose under carbohydrate-limiting conditions. *C. rectus* is also able to produce protohaeme for the growth of black-pigmented anaerobes such as *P. intermedia* and *P. gingivalis*. As can be deduced from this evidence, a complex array of nutritional interactions between bacteria can take place in plaque, with the growth of some species being dependent on the metabolism of other organisms. Indeed the diversity of the plaque microflora is due, in part, to (a) the development of such food chains and food webs (Figure 5.10) and to (b) the lack of a single nutrient limiting the growth of all bacterial species. Bacteria survive by adopting, where possible, alternative metabolic strategies in order to avoid direct competition. Another beneficial interaction among plaque bacteria is co-aggregation; this was described in an earlier section in this chapter.

Antagonism is also a major contributing factor in determining the composition of microbial ecosystems such as dental plaque. The production of antagonistic compounds can give an organism a competitive advantage when interacting with other microbes. One of the most common antagonistic compounds produced are bacteriocins or bacteriocin-like substances. Bacteriocins are relatively high molecular weight proteins that are coded for by a plasmid; the producer strains are resistant to the action of the bacteriocins they produce. Bacteriocins are produced by most species of oral streptococci (e.g. mutacin by *S. mutans* and sanguicin by *S. sanguis*) as well as by *C. matruchotii*, black-pigmented anaerobes, and *A. actinomycetemcomitans*. In contrast, *Actinomyces* species are not generally bacteriocinogenic. Although bacteriocins are usually limited in their spectrum of activity, many of the streptococcal bacteriocins are broad spectrum, inhibiting species belonging to several Gram-positive genera including *Actinomyces* spp. Similarly, a bacteriocin from *S. sanguis* was active not only against most Gram-positive species but also against Gram-negative bacteria including *Capnocytophaga* and *Prevotella* species. Conflicting evidence has been reported on whether such agents would function under the conditions known to exist in plaque. Some reports claim that bacteriocins could be inactivated by proteolytic enzymes in saliva, and by cell-associated EPS, whereas others have found no such evidence. Indeed, gnotobiotic animal studies, in which strains of *S. mutans* with various degrees of bacteriocinogenic activity were implanted along with a bacteriocin-sensitive strain of *A. viscosus*, suggested that bacteriocin-producing bacteria do have an ecological advantage during colonization. Only bacteriocin-producing strains could become established, with the degree of colonization being proportional to the level of *in vitro* activity. Interestingly, even bacteriocin-producing strains of *S. mutans* had difficulty in colonizing when the rat microflora became conventionalized, and hence more complex in composition. The likely explanation for this colonization resistance was that all of the available niches within the microbial community were now occupied.

Other inhibitory factors produced by plaque bacteria include organic acids, hydrogen peroxide, and enzymes. The production of hydrogen peroxide by members of the *S. oralis*-group has been proposed as a mechanism whereby the numbers of periodontopathic bacteria are reduced in plaque to levels at which they are incapable of initiating disease. Perhaps significantly, some periodontal pathogens (e.g. *A.*

actinomycetemcomitans) are able to produce factors inhibitory to oral streptococci. Certain types of periodontal disease (Chapter 7) might result, therefore, from an ecological imbalance between dynamically-interacting groups of bacteria. The low pH generated from carbohydrate metabolism is also inhibitory to many plaque species, particularly Gram-negative organisms and members of the *S. oralis*-group. The production of antagonistic factors will not necessarily lead to the complete exclusion of sensitive species. As discussed previously, the presence of distinct micro-habitats within a biofilm such as plaque enables bacteria to survive that would be incompatible with one another in a homogeneous environment. Also, where there is competition for nutrients, the production of inhibitory factors might be a mechanism whereby less-competitive species can persist (negative feed-back).

Bacterial antagonism will also be a mechanism whereby exogenous (allochthonous) species are prevented from colonizing the oral cavity. The production of inhibitory compounds is normally considered only in terms of interactions between competing resident (autochthonous) organisms but, for example, some *S. salivarius* strains can produce an inhibitor (enocin) with activity against Lancefield Group A streptococci. Enocin-producing strains may prevent colonization of the mouth by this pathogen in a manner similar to that proposed for streptococci in the pharynx. It has been claimed that *S. salivarius* is more frequently isolated from the throats of children who do not become colonized following exposure to Group A streptococci than from those who do become infected. Thus, microbial interactions will play a major role in determining both the final composition and the pattern of development of the plaque microflora.

MICROBIAL HOMEOSTASIS IN DENTAL PLAQUE

In spite of its microbial diversity, the composition of dental plaque at any site is characterized by a remarkable degree of stability or balance among the component species. This stability is maintained in spite of the host defences, and despite the regular exposure of the plaque community to a variety of modest environmental stresses. Strategies by which the resident oral microflora might avoid or evade the host defences were outlined in Chapter 2. Environmental stresses include diet, the regular challenge by exogenous species, the use of dentifrices and mouthwashes containing antimicrobial agents (Chapters 6 and 7), and changes in saliva flow and hormone levels (Figure 5.11). The ability to maintain community stability in a variable environment has been termed microbial homeostasis. This stability stems not from any metabolic indifference among the components of the microflora but results

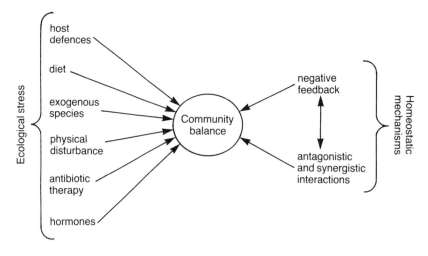

Figure 5.11 Factors involved in the maintenance of microbial homeostasis in the mouth.

rather from a balance of dynamic microbial interactions, including both synergism (such as proto-co-operation and commensalism) and antagonism. When the environment is perturbed, self-regulatory mechanisms (homeostatic reactions) come into force to restore the original balance. An essential component of such mechanisms is negative feedback, whereby a change in one or more organisms results in a response by others to oppose or neutralize such a change. There is a tendency for homeostasis to be greater in microbial communities with an increased species diversity.

Despite the microbial diversity of dental plaque, homeostasis does break down on occasions. The main causes for this can be divided into either (a) deficiencies in the immune response, or (b) other (non-

Table 5.9 Factors responsible for the breakdown of microbial homeostasis

Immunological factors	Non-immunological factors
sIgA-deficiency	Xerostomia
Neutrophil dysfunction	Antibiotics
Chemotherapy-induced myelosuppression	Dietary carbohydrates/low pH
	Increased GCF flow
Infection (AIDS)	Oral contraceptives

immune) factors (Table 5.9). The host defences, together with the resident microflora, serve to maintain microbial homeostasis in plaque (and on other oral surfaces), and together they act synergistically to prevent colonization by exogenous species and the invasion of host tissues by opportunistic pathogens. The remainder of this book will be devoted to describing the consequences of the breakdown of microbial homeostasis in the mouth, and will describe the aetiology of the major oral diseases.

SUMMARY

Dental plaque is the microbial community, embedded in polymers of salivary and bacterial origin, found on the tooth surface. The development of dental plaque is an example of autogenic succession whereby microbial factors influence the pattern of the development of the microflora. The formation of dental plaque can be divided arbitrarily into a number of distinct stages. These include a reversible phase involving van der Waals attractive forces and electrostatic repulsion, an irreversible phase involving specific inter-molecular interactions between bacterial adhesins and host ligands, coaggregation of bacteria to already-attached organisms, and cell division leading to confluent growth and biofilm formation. The pioneer species include members of the S. oralis-group, haemophili, and *Neisseria* species. These organisms do not colonize clean enamel but interact with a layer of proteins and glycoproteins (derived mainly from saliva) adsorbed onto the tooth surface (the acquired pellicle). The process of plaque formation leads to an increase in the diversity of the microflora, although the composition of the climax community varies at different sites on the tooth surface. The microbial community of fissures is less diverse than that of approximal sites and the gingival crevice. Obligately anaerobic bacteria form a significant part of the microflora from these latter two sites, so special precautions are necessary when sampling and processing plaque from these areas in order to maintain the viability of the resident micro-organisms. Variations occur in the plaque microflora at the same site both between and within mouths; variations can also occur at the same site over relatively small distances, so that small samples are necessary if important site variations are not to be overlooked.

The balance of the microflora at a site remains reasonably stable unless severely perturbed by an environmental stress. Such a stable microflora is also able to prevent exogenous species from colonizing. This stability (termed microbial homeostasis) is due, in part, to a dynamic balance of microbial interactions, including synergism and antagonism. Synergistic interactions include coaggregation, the development of food-chains, and the degradation of complex host and bacterial polymers. Antagonism can be due to the production of bacteriocins, hydrogen peroxide, enzymes, organic acids and low pH. The spatial heterogeneity of a biofilm such as plaque can lead to the co-existence of species that would be incompatible with one another in a homogeneous environment. Dental plaque must never be regarded as a constant, static ecosystem: a consideration of the points raised throughout this chapter serve to emphasize its dynamic nature.

FURTHER READING

Bowden, G. H., Hardie, J. M. and Slack, G. L. (1975) Microbial variations in approximal dental plaque. *Caries Research*, **9**, 253–77.

Edgar, W. M. and Higham, S. M. (1990) Plaque fluid as a bacterial milieu. *Journal of Dental Research*, **69**, 1332–6.

Gibbons, R. J. (1989) Bacterial adhesion to oral tissues: a model for infectious diseases. *Journal of Dental Research*, **68**, 750–60.

Kolenbrander, P. E. (1988) Intergeneric coaggregation among human oral bacteria and ecology of dental plaque. *Annual Reviews of Microbiology*, **42**, 627–56.

Liljemark, W. F., Fenner, L. J. and Bloomquist, C. G. (1986) *In vivo* colonization of salivary pellicle by *Haemophilus*, *Actinomyces* and *Streptococcus* species. *Caries Research*, **20**, 481–97.

Marsh, P. D. (1989) Host defences and microbial homeostasis: Role of microbial interactions. *Journal of Dental Research*, **68**, 1567–75.

Mergenhagen, S. E. and Rosan, B. (eds.) (1985) *Molecular Basis of Oral Microbial Adhesion*. American Society for Microbiology, Washington DC.

Nyvad, B. and Kilian, M. (1987) Microbiology of the early colonization of human enamel and root surfaces *in vivo*. *Scandinavian Journal of Dental Research*, **95**, 369–80.

Slots, J. (1977) Microflora of the healthy gingival sulcus in man. *Scandinavian Journal of Dental Research*, **85**, 247–54.

Theilade, E., Budtz-Jorgensen, E. and Theilade, J. (1983) Predominant cultivable microflora of plaque on removable dentures in patients with healthy oral mucosa. *Archives of Oral Biology*, **28**, 675–80.

Theilade, E., Fejerskov, O., Karring, T. and Theilade, J. (1982) Predominant cultivable microflora of human dental fissure plaque. *Infection and Immunity*, **36**, 977–82.

6 Dental caries

In industrialized societies, enamel caries affects the vast majority of individuals, particularly up to the age of 20 years, whereafter its incidence is reduced. Root-surface caries is becoming a problem in the elderly due to gingival recession exposing the vulnerable cementum to microbial colonization. Dental caries can be defined as the localized destruction of the tissues of the tooth by bacterial action. Cavities begin as small demineralized areas on the surface of the enamel, and can progress through the dentine and into the pulp. Demineralization of the enamel is caused by acids, and in particular lactic acid, produced from the microbial fermentation of dietary carbohydrates. Lesion formation involves dissolution of the enamel and the transport of the calcium and phosphate ions away into the surrounding environment. This initial stage is reversible and remineralization can occur, particularly in the presence of fluoride.

SPECIFIC AND NON-SPECIFIC PLAQUE HYPOTHESES

Preventive and curative regimens for caries, periodontal disease and other diseases would be more precise if the particular micro-organism causing the disease was known. For any microbe to be considered responsible for a given condition, Koch's postulates have been applied, and these are:

1. The microbe should be found in all cases of the disease with a distribution corresponding to the observed lesions.
2. The microbe should be grown on artificial media for several subcultures.
3. A pure subculture should produce the disease in a susceptible animal.

To these three Koch's postulates is added:

4. A high antibody titre to the microbe should be detected during infection, this may provide protection on subsequent reinfection.

Despite extensive sampling of 'periodontal' and 'carious plaques' together with gnotobiotic and germ-free animal experiments, no single microbe has been found which satisfies Koch's postulates for caries, periodontal disease or any other opportunistic oral infection. Koch's postulates also imply that infections reach homeostasis with the microbial content becoming constant and a pathogen becoming predominant. It has become apparent in recent years that all infections have a progression (natural history) which has evolution and possible stasis. Even when a microbe eventually predominates it is usually interdependent on others present in the lesion to maintain it. Thus, for any given infection there may be an apparently heterogeneous group of causative micro-organisms. This has important implications in the diagnosis of opportunistic infections, or in the study of putative pathogens in caries and periodontal diseases. In the past, numerical predominance of a micro-organism isolated from a lesion has been the guiding principle in the diagnosis of infection; this utilized Koch's postulates. This diagnostic principle is now difficult to support as the numerical predominance of a microbe takes no account of the contribution that it makes to the maintenance or progression of the lesion. It has therefore been suggested that Koch's postulates should be modified for opportunistic infections in the following way:

1. A microbe should be present in sufficient numbers to initiate disease.
2. The microbe should have access to the affected tissues.
3. The microbe should be in an environment that permits its survival and multiplication.
4. Other micro-organisms that inhibit the growth of the microbe should be absent or, if present, should not significantly affect it.
5. The host must be 'susceptible' to the microbe.

On the basis of these principles it is more accepted in diagnostic oral microbiology to report that two or three micro-organisms may be responsible for an infection. This sometimes alters antibiotic treatment where the susceptibilities of several micro-organisms must be considered if therapy is to be successful.

Caries and periodontal disease are both plaque-associated infections. The types of micro-organisms found in plaque associated with disease will be discussed in this Chapter and in Chapter 7. There are two main schools of thought on the role of plaque bacteria in the aetiology of these two groups of disease. The **Specific Plaque Hypothesis** proposes that, out of the diverse collection of organisms comprising the resident plaque microflora, only a limited number of species are involved in disease. In contrast, the **Non-Specific Plaque Hypothesis** considers that any of a heterogeneous mixture of micro-organisms can play a role in disease aetiology. In some respects, the arguments may be about semantics, since plaque-mediated diseases are essentially mixed culture infections, but in which certain, perhaps specific, species predominate. The arguments then centre around the definitions of the terms specific and non-specific. If not actually specific, then the diseases certainly show evidence of specificity.

EVIDENCE FOR CARIES AS AN INFECTIOUS DISEASE

In the last century, Leber and Rottenstein in 1867 and Miller in 1890 deduced the fundamental principles involved in dental caries. In his famous Chemico-Parasitic Theory, Miller suggested that oral bacteria converted dietary carbohydrates into acid which solubilized the calcium phosphate of the enamel to produce a caries lesion. Although Clarke isolated an organism (which he called *Streptococcus mutans*) from a human caries lesion in 1924, proof for the causative role of bacteria came only in the 1950s and 1960s following experiments with germ-free animals.

Pioneer experiments showed that germ-free rats developed caries when infected with bacteria described as enterococci. Evidence for the transmissability of caries came from studies on hamsters. Caries-inactive animals had no caries even when fed a highly cariogenic (i.e. sucrose-rich) diet. Caries only developed in these animals when they were caged with or ate the faecal pellets of a group of caries-active hamsters. Further proof came when streptococci, isolated from caries lesions in rodents, caused rampant decay when inoculated into the oral cavity of previously caries-inactive hamsters. The importance of diet became apparent when the colonization and production of caries by most streptococcal populations occurred only in the presence of sucrose. Subsequent research has shown that some oral streptococci not only produce acid from sucrose (i.e. they are acidogenic), but also they can tolerate the low pH so-produced (i.e. they are also aciduric), and synthesize extracellular polysaccharides that are important in oral colonization and plaque development (Chapters 4 and 5).

Mutans streptococci can cause caries of smooth surfaces, as well as in pits and fissures, in hamsters, gerbils, rats and monkeys fed on cariogenic diets, and are the most cariogenic group of bacteria found. Other bacteria, including members of the *S. oralis-*, '*S. milleri-*' and *S. salivarius-* groups, *E. faecalis*, *A. naeslundii*, *A. viscosus* and lactobacilli can also produce caries under conducive conditions in some animals, although the lesions are usually restricted to fissures. Evidence for the significance of the role of mutans streptococci in dental caries has also come from vaccination studies. Immunization of rodents or monkeys with whole cells or specific antigens of *S. mutans* and *S. sobrinus* leads to a reduction in the number of these organisms in plaque and a decrease in the number of caries lesions compared with sham-immunized animals.

AETIOLOGY OF HUMAN ENAMEL CARIES

Unlike the studies of animals, any relationship between particular oral bacteria and caries in humans must be derived by indirect means. Evidence for bacterial involvement has come from several sources. Patients on long-term broad-spectrum antibiotic therapy frequently exhibit a reduced caries experience. Similar results are found with experimental animals kept on diets supplemented with antibiotics active against Gram-positive species. A variety of epidemiological surveys of different human populations have found a strong association between mutans streptococci and caries. Much research effort over the past two decades has been focused on determining the precise bacterial aetiology of caries so that effective preventive measures can be devised. The findings from these studies will be reviewed and summarized in the following sections.

Natural history of dental caries

Cavitation is the final stage of enamel caries; it is preceded by a clinically-detectable small lesion, known as a 'white spot', and before that by sub-surface demineralization, which can only be detected by histological techniques. Caries often occurs in teeth shortly after eruption (hence, its association with young people), and some teeth and surfaces are more vulnerable than others. The prevalence of caries is highest on the occlusal surfaces of first and second molars, and lowest on the lingual surfaces of mandibular teeth. The risk to approximal surfaces is intermediate to those described above. Some individuals are more caries prone than others, and this may be related to a higher frequency of sugar intake or to a severely reduced saliva flow (for example, as in xerostomic patients). The microbiological findings have tended to be more obviously different between caries active and inactive populations.

Not all white-spot lesions progress to cavitation; in some studies only about half of these early lesions penetrated the dentine after 3–4 years. The incidence of progression is even lower on smooth surfaces; occlusal lesions, however, tend to progress more often and more rapidly. White-spot lesions do not just arrest, but can even remineralize; remineralization is enhanced by fluoride.

Implications for the design of studies of the aetiology of caries

This behaviour of lesions to de- and remineralize at different rates can complicate the interpretation of clinical studies. In cross-sectional surveys, where cavitation is diagnosed, it cannot be certain whether the species that are isolated caused the decay or arose because of it. Likewise, it cannot be determined whether the lesion, at the time of sampling, was progressing, arrested, or healing, and each phase may have a different microflora. To overcome these difficulties, longitudinal studies have been designed in which clinically-sound sites are sampled at regular intervals over a set time period. Surfaces are chosen so that a reasonable proportion will decay within the time span of the study. The microflora can then be compared: (a) before and after the diagnosis of a lesion, and (b) between those surfaces that decayed and those that remained caries-free throughout the study. A disadvantage of this approach is the fact that plaque analysis is time-consuming and laborious (Chapter 4), so that only a limited number of individuals can be followed. Consequently, relatively few longitudinal studies, in which true cause-and-effect relationships can be established, have been performed, compared to the numerous cross-sectional surveys that have been reported. Only 'associations' can be derived from these latter study designs, but they have the advantage that large numbers of sites/people can be analysed, and different patient groups, age groups, tooth surfaces, etc. can be screened quickly.

MICROBIOLOGY OF ENAMEL CARIES

Superimposed on the problems of study design outlined above, are those associated with the microbiological analysis of plaque. Many of these were outlined in Chapter 4. The plaque microflora is diverse, and disease is not due to exogenous species, which would be easy to identify, but to changes in the relative proportions of members of the resident microflora. There are wide inter-subject variations in the composition of the plaque microflora (Chapter 5), so that when data is meaned from numerous individuals, clear associations between bacteria and disease can be obscured. Likewise, once mutans streptococci had been strongly implicated in human caries, many follow-up studies only looked to confirm this association. In such studies, the role of any other species could not be determined. Despite these problems of study design and methodology, much progress has been made, and the major findings will now be discussed.

Smooth surface caries

Buccal and lingual smooth surfaces are easy to clean and hence, suffer from decay only rarely. However, they are easy to study for experimental purposes, both in terms of clinical diagnosis and in plaque sample taking. Early studies showed higher proportions of mutans streptococci on white-spot lesions on smooth surfaces compared to sound enamel. Subsequent studies found that the levels of mutans streptococci in a white-spot lesion were frequently 10–100 fold higher than on immediately-adjacent sound enamel (Table 6.1). However, as stated earlier, such an association does not prove a causal relationship. Except for one site, the actual proportions of mutans streptococci were low. The bacteria making up the remainder of the plaque microflora were not identified but would include *Actinomyces* spp. and other species of streptococci, which can also make significant concentrations of acid from carbohydrates.

Table 6.1 The percentage viable count of mutans streptococci (MS) from a white spot lesion and from adjacent sound enamel on the same tooth

Subject	Mean percentage viable count of mutans streptococci		Ratio MS white spot/ MS sound enamel
	White spot lesion	*Sound enamel*	
1	1.30	0.03	43.3
2	0.01	0.002	5.0
3	0.06	0.002	30.0
4	0.06	0.02	3.0
5	0.8	0.01	80.0
6	63.2	0.5	126.4
7	1.4	0.2	7.0
8	0.2	0.07	0.3

The viable counts are the mean value of several sites in the white spot lesion and an equal number of sites on neighbouring sound enamel. Eight subjects were studied.

Approximal surface caries

A problem with studies aimed at determining the microbial aetiology of caries at approximal sites is that early lesions cannot be diagnosed accurately, and plaque samples are inevitably removed from the whole interproximal area. This will include enamel that is clinically sound as well as that undergoing demineralization. The microflora can vary markedly at different sites around the contact area (Table 5.5), irrespective of whether a lesion is developing, so that specific associations can be obscured. In general, the data concerning the specificity of particular species and caries initiation has been more equivocal at approximal sites than at other surfaces. Much of this may be due to the technical difficulties of lesion sampling and diagnosis outlined above.

Early cross-sectional studies found a positive correlation between elevated levels of mutans streptococci and lesion development. This has been confirmed in other studies which have shown that more lesions develop on surfaces that are colonized by mutans streptococci than those that are not. Furthermore, the likelihood of caries rises with increased levels of mutans streptococci in plaque. Many of these studies, however, are limited in scope, and only monitor a small number of micro-organisms, and sometimes only mutans streptococci. In order to combat this, a major longitudinal study was carried out on English schoolchildren, aged 11–15 years, in which all of the predominant plaque microflora was characterized. Not surprisingly, the enormous amount of bacteriological and clinical data generated from this study caused problems in analysis, but no unique association between any organism and caries initiation was found. Mutans streptococci could be found in high numbers prior to demineralization at a number of sites, but lesions also appeared to develop in the apparent absence of this group of bacteria; these organisms could also be present at other sites in equally high numbers without any evidence of caries. The latter situation might arise in sites where lactate-utilizing bacteria or base-generating species are highly prevalent, or it may be due to differences in the activity of host factors, such as the rate of flow, the composition, or buffering capacity of saliva. Some of the microbiological data from this study are shown in Figure 6.1. It appeared that both the isolation frequency and proportions of mutans streptococci increased after, rather than before, the radiographic detection of a lesion. Similar findings were found with lactobacilli, and some of these early lesions progressed deeper into the enamel. This provided some of the earliest evidence that shifts in the composition of the microflora might occur as the lesion progresses through the tissues of the tooth.

Analogous findings were found in a study of Dutch army recruits,

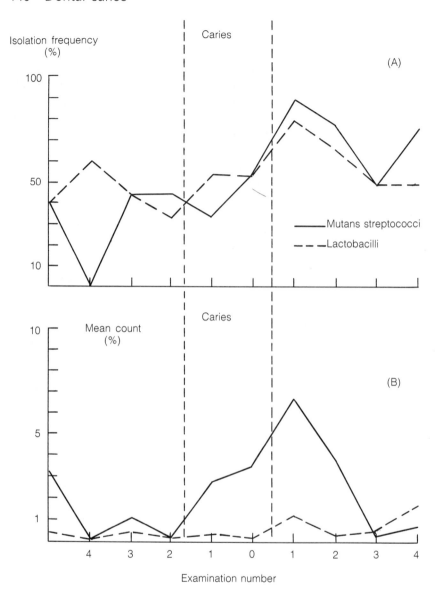

Figure 6.1 Isolation frequency (A) and mean percentage viable count (B) of mutans streptococci and lactobacilli in plaque samples from approximal surfaces of 15 children before and after the initial detection of a caries lesion. Time 0 = examinations when lesions were first detected radiographically; time periods on the left = examinations prior to lesion detection; time periods on the right = examinations after lesion detection.

aged 18–20 years. Mutans streptococci were isolated from 40% and 86% of sites from caries-free and caries-active recruits, respectively. Interestingly, marked differences in the distribution of individual species were found; *S. mutans* strains were isolated from both groups, whereas *S. sobrinus* was recovered almost exclusively from caries-active recruits. The prevalence of the combined species of mutans streptococci showed a direct correlation with the progression of a lesion into the dentine. Caries progression occurred on 14 tooth surfaces, 71% of which harboured percentages of mutans streptococci greater than 5% of the total cultivable plaque microflora, while this species was detected only on 51% of the tooth surfaces without caries progression (Table 6.2). Again, the association of mutans streptococci with approximal caries was not absolute. Relatively high proportions of mutans streptococci persisted at tooth surfaces without caries progression while caries could also develop in their apparent absence.

Table 6.2 Prevalence of mutans streptococci at approximal tooth surfaces with and without caries progression

	Total number of tooth surfaces	Prevalence of mutans streptococci*:		
		0%	0–5%	>5%
Caries progression	14	1†	3	10
No caries progression	41	21	17	3

* Prevalence of mutans streptococci is expressed as their percentage of the cultivable microflora
† Number of tooth surfaces.

Fissure caries

Fissures on occlusal surfaces (Figures 2.1 and 5.7) are still the most caries-prone sites on the teeth. Caries can develop rapidly on these surfaces, and it is at these sites that the strongest association between mutans streptococci and dental decay has been found. In one cross-sectional study, 71% of carious fissures had viable counts of mutans streptococci >10% of the total cultivable plaque microflora, whereas 70% of the fissures that were caries free at the time of sampling had no detectable mutans streptococci. An inverse relationship between mutans streptococci and *S. sanguis* is frequently observed. Many other studies on different population groups around the world (but predominantly in industrialized societies) have confirmed these trends.

Similar conclusions have been drawn from longitudinal studies. The proportions of mutans streptococci, *S. sanguis*, and lactobacilli were monitored before and at the time of caries development on occlusal fissures in American children. The subjects were divided into several groups according to their previous caries experience and to their caries activity during the study; these groups are defined in Table 6.3. A longitudinal analysis of the data showed that the percentage count of mutans streptococci increased significantly at the time of caries diagnosis of most of the occlusal lesions. Cross-sectional comparisons also demonstrated significantly higher proportions of mutans streptococci in the fissures that became carious than in those that remained caries-free (Table 6.3). However, mutans streptococci were only a minor component of the plaque from five fissures which became carious. Counts of lactobacilli were significantly higher at these sites and, it was concluded, that it was these organisms that were responsible for lesion formation at these sites. Furthermore, in a group who remained caries-free during the study, but who had previously a high caries experience (HCI group, Table 6.3), mutans streptococci comprised, on average, approximately 8% of the total cultivable microflora over a 12-month period. Thus, although there was a strong correlation between mutans streptococci and fissure decay, lesions could also develop in the absence of this group of bacteria, and these bacteria could persist in moderately high numbers at apparently caries-free sites. In a more recent longitudinal study, it was proposed that once the level of mutans streptococci in a fissure reached 20% of the total cultivable microflora, then there was a high probability of a detectable lesion developing within the following six months.

It is possible that the failure to recover mutans streptococci from fissures with caries might be due to sampling difficulties. Most techniques using needles or toothpicks only remove a fraction of the total microflora (approximately 20%), and might not detect mutans streptococci present in the depth of the fissure. It is noteworthy that organisms such as lactobacilli appear to preferentially inhabit the lower depths of the fissure. Studies of the distribution of mutans streptococci have isolated *S. mutans* more commonly than *S. sobrinus* from occlusal surfaces, whereas *S. sobrinus* was found more frequently on molars than anterior teeth.

Table 6.3 Percentage viable count of mutans streptococci, S. *sanguis*, and lactobacilli in occlusal fissure plaque of children participating in a longitudinal study relating plaque microflora to caries initiation

Carious status of tooth		Percentage viable count		
	No. of teeth	mutans streptococci	S. sanguis	Lactobacilli
Carious				
LCA	5	0.1	4.6	4.0
HCA	37	24.6	7.8	1.0
Caries-free				
CF	24	2.0	12.8	0.3
LCI	31	1.6	20.6	0
HCI	24	8.3	6.4	0.2
LCA	6	0	7.0	0
HCA	68	7.2	12.3	0.4

LCA caries diagnosed at target site during study, but with a previously low experience of caries

HCA caries diagnosed at target site during study, but with a previously high experience of caries

CF caries free

LCI caries inactive (no new lesions developed during the study) but with a previously low experience of caries

HCI caries inactive (no new lesions developed during the study) but with a previously high experience of caries.

Rampant caries

Rampant caries can occur in particular sub-groups of people who are especially prone to decay for specific reasons. As stated earlier, one group is the xerostomic patients who have a markedly reduced salivary flow rate, for example, due to radiation treatment for head and neck cancer, to Sjögrens syndrome, or to medication. Not only do these patients have severely reduced salivary flow rates, but they generally consume soft diets, with a high sucrose content, and may often suck 'candies' to relieve their symptoms. Longitudinal studies of patients undergoing radiation treatment showed large increases in the numbers and proportions of mutans streptococci and lactobacilli in plaque and saliva. Other species, associated with sound enamel, such as S. *sanguis*, *Neisseria* and Gram-negative anaerobes, decreased during this period.

Analysis of these findings also suggested that there may be waves of bacterial succession occurring, with mutans streptococci being associated with caries initiation in these patients while lactobacilli may be involved with lesion progression.

Rampant caries is also found in another specific situation and that is 'nursing-bottle' caries. This is the extensive and rapid decay of the maxillary anterior teeth associated with the prolonged and frequent feeding of young infants with bottles or pacifiers containing formulas with a high concentration of fermentable carbohydrate. Plaque bacteria receive an almost continuous provision of substrates from which they can make acid. Such prolonged conditions of low pH are conducive, and indeed, selective for mutans streptococci and lactobacilli. Not surprisingly, therefore, some of the highest proportions (>50%) of mutans streptococci in plaque have been reported from sites affected by this condition. Lactobacilli (including *L. fermentum* and *L. plantarum*) are also prevalent in plaque with this type of caries. In one Canadian study, however, it was noted that similar microfloras could be recovered from caries-free susceptible sites and caries lesions, implying that local factors could influence the cariogenicity of the plaque.

Early (sub-surface) demineralization

As discussed earlier, a problem with many studies of caries is that diagnosis of a lesion has to rely on relatively insensitive techniques such as radiographs or tactile criteria. By the time a lesion can be reliably diagnosed in this way, it is relatively well-advanced. It is possible that the bacteria associated with the established lesion may not be the same as those responsible for its initiation. In order to circumvent these problems, two studies have examined the integrity of enamel from teeth that were being extracted for orthodontic purposes. Very small samples of plaque could be taken from discrete sites around the contact area on approximal surfaces; demineralization could be determined by sensitive histological criteria using polarized light microscopy or microradiography. One study used orthodontic bands as an *in vivo* model system to create artificially a protected area for plaque accumulation. Surprisingly, wide differences in demineralization rates were found. Demineralization could occur in some individuals even after as little as two days of plaque accumulation, while it could be questionable in other subjects even after 14 days. Lactobacilli, *S. mutans* and *A. viscosus* were detected more commonly at day 14 than at day 1. Demineralization was more common in the presence of *S. mutans*, although demineralization also occurred in 24% of sites from which *S. mutans* could not be isolated.

Similar trends were found in the second study; the proportions of mutans streptococci and *A. viscosus* were found to be statistically higher at sites with demineralization. Very early stages of demineralization were also associated with increases in *A. odontolyticus*; in contrast, lactobacilli were only isolated from the later stages of demineralization. Again, mutans streptococci could not be detected from 37% of sites with pronounced demineralization. It should be borne in mind, however, that this type of study suffers from the problem discussed earlier that it is not possible to ascertain whether the demineralization was active, arrested or even repairing at the time of sampling. Collectively, however, the two studies emphasize the role of mutans streptococci in early demineralization; they also point to a potential role for *A. viscosus*, with lactobacilli becoming prevalent only at the later stages of lesion formation.

Lesion progression

Indirect evidence from many of the preceding sections has suggested that bacterial succession may occur during the development of a caries lesion, and that the microflora responsible for initiation may differ from that causing progression. Data to support this contention has come from a 12-month longitudinal study of the plaque microflora of small incipient approximal lesions in deciduous molar teeth of schoolchildren (aged 4–9 years). During this period, some of the lesions progressed while others did not. Differences were found in the plaque microflora at the start of the study between lesions which ultimately progressed and those that were arrested (Table 6.4). Lactobacilli and *A. odontolyticus* were isolated in low numbers (<1% of the total microflora), but only from lesions which eventually progressed, and were never isolated from non-progressive lesions, or from sound enamel. Other positive associations with caries progression were found with mutans streptococci and *Veillonella* spp.; organisms showing a negative correlation included members of the *S. oralis*-group and *A. naeslundii* (Table 6.4). These findings might find clinical application in that the detection of lactobacilli or *A. odontolyticus* at incipient lesions would indicate a site with a high-risk of cavitation. Such a finding, therefore, could warrant the topical application of agents to either encourage remineralization (e.g. fluoride) or disrupt the ecology of the microflora (e.g. antimicrobial agents).

Table 6.4 Prevalence and isolation frequency of some bacteria recovered from progressive and non-progressive incipient caries lesions in young schoolchildren

Bacterium	Progressive lesion		Non-progressive lesion	
	%age viable count	%age isolation frequency	%age viable count	%age isolation frequency
mutans streptococci	10	100	3	75
S. sanguis-group	6	100	9	87
S. oralis-group	4	50	5	100
A. viscosus	29	100	33	100
A. naeslundii	<1	50	5	50
A. odontolyticus	<1	50	0	0
Lactobacillus spp.	<1	75	0	0
Veillonella spp.	11	100	9	100

Recurrent caries

Caries can re-occur beneath and around previous restorations. This might be due to penetration of micro-organisms around the margins of poorly-fitting restorations or due to the incomplete removal of bacteria when the initial lesion was originally restored. There have been few studies of this problem, but mutans streptococci have been isolated in high numbers from recurrent caries while lactobacilli have also been isolated when dentine is affected. Interestingly, the type of restoration may influence the development of the microflora and affect recurrent caries. Conventional amalgam is being replaced in some situations by new materials including glass ionomer cement. This material can leach fluoride and silver ions into the immediate environment; both ions can exert an antibacterial effect. Plaque removed from approximal sites restored with glass ionomer cement had lower levels of mutans streptococci than those restored with amalgam.

MICROBIOLOGY OF ROOT-SURFACE CARIES

Unlike enamel caries, root-surface caries is a disease of middle-aged and older adults. Gingival recession exposes the cementum of the root to microbial colonization; root surfaces can also become exposed due to mechanical injury or to periodontal surgery (e.g. scaling and root plan-ing; Chapter 7). These cementum surfaces, when colonized by bacteria, are especially vulnerable to demineralization by acid, and therefore, to caries. Root surface caries was common in ancient populations and is prevalent in present day primitive communities. In contemporary populations, the prevalence of root surface caries increases with age. This is now becoming a major problem due to people retaining their teeth for longer periods. Approximately 60% of individuals aged 60 years or older now have root caries or fillings.

Experimental animal studies

Direct evidence for the role of oral micro-organisms in root-surface caries came from animal studies in the 1960s in which filamentous bacteria were found to invade the root surfaces of hamsters and produce caries. Human isolates of *A. naeslundii* and *A. viscosus* were then shown to cause root surface caries (and periodontal disease) in gnotobiotic rats and hamsters, as were pure cultures of mutans streptococci, *S. mitis*, 'S. milleri' and *S. sanguis* in subsequent experiments; studies with *A. israelii* have given equivocal results, while lactobacilli only caused enamel caries in animals.

Human studies

Early studies were designed around the findings from the first animal experiments, and focused on the role of Gram-positive filamentous bacteria, especially *Actinomyces* spp. in root surface caries. Among the organisms isolated from lesions were *Rothia dentocariosa*, *A. viscosus*, *A. naeslundii* and *A. odontolyticus*; in some studies mutans streptococci were also reported to be significant components of the microflora. Indeed, some lesions could be distinguished by those in which mutans streptococci predominated, and those in which they were absent and *Actinomyces* spp. (especially *A. viscosus*) were the dominant organisms.

More recent studies have tended to find stronger associations between mutans streptococci and lactobacilli with root surface caries. In a major longitudinal survey in Canada, although no specific correlations between certain bacteria and root caries were found, the presence of mutans streptococci and lactobacilli on root surfaces were predictive for the subsequent development of a lesion. Other studies have attempted to sub-divide the lesions into initial (or 'soft') and advanced (or 'hard'). Several groups reported higher proportions (often around 30% of the total cultivable microflora) of mutans streptococci at the initial lesion (Table 6.5), sometimes in association with lactobacilli. Advanced lesions did not appear to be associated with a specific group of micro-organisms, but the microflora was variable and diverse, with Gram-positive rods such as *Actinomyces* spp., *Propionibacterium* (formerly *Arachnia*) *Bifidobacterium*, *Lactobacillus* and *Rothia* being isolated by different groups. The collective data suggest that bacterial succession also occurs during the development of root surface lesions.

Table 6.5 Mean percentage viable counts (and percentage isolation frequences in parentheses) of some plaque bacteria from root surfaces, with and without caries

| | | Root surface caries | |
Bacterium	Sound root surface	Initial (soft)	Advanced (hard)
Mutans streptococci	2 (84)	29 (92)	8 (92)
S. sanguis	19 (96)	11 (97)	22 (85)
A. viscosus/naeslundii	12 (90)	11 (85)	13 (96)

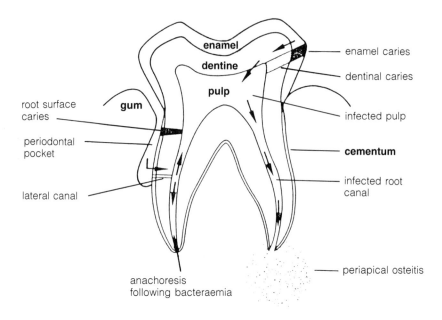

Figure 6.2 Progression of infections affecting the tooth and its supporting structures.

BACTERIAL INVASION OF DENTINE AND ROOT CANALS

Dentine can be invaded by (a) direct progression of an enamel caries lesion; (b) from caries of the root surface; (c) from a periodontal pocket via lateral or accessory canals (Figure 6.2), or (d) as a result of fracture or trauma during operative procedures. Scaling and root planing can predispose some root surfaces to bacterial invasion by exposing dentine tubules; in addition, bacteria may lodge in injured pulp following a transient bacteraemia (anachoresis) (Figure 6.2).

The microbial community from the advancing front of a dentinal lesion is diverse and contains many obligately anaerobic bacteria belonging to the genera *Actinomyces*, *Bifidobacterium*, *Eubacterium*, *Lactobacillus*, *Propionibacterium* and *Rothia*. Gram-negative species can also be isolated but they are generally present only in low numbers. Streptococci are recovered less frequently, but when mutans streptococci have been isolated they can be one of the predominant members of the community.

Bacteria have also been demonstrated by microscopy in the intact radicular cementum and dentine of periodontally-diseased (but caries-free) teeth of man and animals. The bacteria from these sites have not usually been identified, but both motile and non-motile species have been observed. *In vitro* experiments have shown that mutans streptococci, *A. naeslundii* and *C. gingivalis* have the potential to invade dentinal tubules. Recently, it was shown that the microflora found in the dentine and pulp of periodontally-diseased human teeth were derived predominantly from the subgingival area. Both Gram-positive and Gram-negative species were identified (Table 6.6); some were more prevalent in the dentine (e.g. *A. viscosus*), some predominated in the pulp (e.g. black-pigmented anaerobes), while others were found equally at both sites (e.g. *A. naeslundii*, *Veillonella* spp., *F. nucleatum*). The microflora was highly variable and generally consisted of diverse mixtures of bacteria; some species were found only once, while others were more commonly isolated. Little is known about the nutritional sources at these sites, but both saccharolytic and asaccharolytic species appear to flourish.

Once bacteria are in the pulp, inflammation can occur which may result eventually in necrosis of the root canal. A further consequence is that micro-organisms can invade and destroy tissue surrounding the apex of the root, producing a spreading or localized infection (periapical osteitis; Figure 6.2). A diverse mixed culture of bacteria can be cultured, including black-pigmented anaerobes, *Campylobacter sputorum*, *Eubacterium* spp. and anaerobic streptococci. The black-pigmented anaerobes include *P. intermedia*, *P. melaninogenica*, and *Porphyromonas endodontalis*.

Table 6.6 Prevalence of some bacteria invading the dentine and pulp of periodontally-diseased teeth

Bacterium	Mean %age viable count*		%age isolation frequency	
	Dentine†	Pulp	Dentine†	Pulp
S. sanguis	46	56	28	7
S. intermedius	10	56	3	2
Peptococcus spp.	15	20	5	2
Peptostreptococcus spp.	2	0	2	0
A. viscosus	47	0	17	0
A. naeslundii	57	61	24	13
A. odontolyticus	40	2	11	3
Bifidobacterium spp.	18	0	3	0
Eubacterium spp.	54	66	3	7
Lactobacillus spp.	1	0	2	0
P. propionicus	21	0	2	0
Propionibacterium spp.	36	22	23	2
Clostridium spp.	37	0	12	0
Veillonella spp.	29	34	7	8
Black-pigmented anaerobes	8	20	13	7
Capnocytophaga spp.	19	3	20	7
F. nucleatum	32	37	15	5
S. sputigena	0	11	0	2
E. corrodens	2	0	2	0

* mean value in positive samples
† mean value for inner, middle and outer layers of dentine.

The latter organism is found almost exclusively in infected root canals and abscesses of endodontal origin; *P. gingivalis* has also been isolated from endodontal abscesses. *Mitsuokella dentalis* is a recently described species that is also strongly associated with infected root canals.

The treatment of infections of the root canal (endodontics) involves the removal of infected and dead tissue both mechanically and by irrigation, sometimes accompanied by treatment with antimicrobial agents to reduce the microbial community to a level where the cavity can be restored effectively. Conventionally, after treatment with antimicrobial agents, each canal is sampled for the presence of micro-organisms. When samples no longer contain viable cells, the canal is deemed sterile and is restored. The validity of this approach has been questioned. It is impossible to be certain of the complete eradication of bacteria from all the tubules radiating from the pulp. Indeed, only the most rigorous sampling and sophisticated culturing will recover many of these bacterial populations and such methods are not routinely available. Paper points cultured in broth are of dubious value since oral contaminants will result in rapid overgrowth of potentially important populations, and no estimate of the concentration of a particular species can be made. It has been argued that the culturing of each canal is unnecessary. Proven endodontic procedures should be adopted which can be relied upon to produce the desired result. While detailed studies of the root microflora would be important in both the design and initial monitoring of such procedures, subsequent microbiology should be necessary only for root canals not responding to treatment. However, a strong case can be made for the routine culturing of root canals by undergraduates as an indication of the success of their aseptic techniques.

PATHOGENIC DETERMINANTS OF CARIOGENIC BACTERIA

Dental caries results from the interaction of the plaque microflora and the diet of the host with a susceptible enamel or cemental surface. The susceptibility of the surface can be modified by a number of factors including the flow rate and buffering capacity of saliva, the activity of the innate and specific host defence systems (Chapter 2), and the presence of fluoride. Much attention has been paid to the identification of determinants of pathogenicity of cariogenic bacteria, such as mutans streptococci and lactobacilli (Table 6.7). Mutants of *S. mutans* that are deficient in either extracellular (EPS) or intracellular polysaccharide (IPS) synthesis form less plaque and produce fewer caries lesions than parent strains when inoculated in pure cultures in gnotobiotic rodents. However, these properties are only partially responsible for virulence because other non-cariogenic streptococci also produce insoluble EPS (e.g. *S. sanguis*), while naturally low IPS-synthesizing species, such as *S. sobrinus*, have been associated with human caries. Moreover, lactobacilli do not synthesize EPS or IPS and yet are implicated with the progression of lesions. Two features that, if not unique, are certainly distinctive properties of cariogenic bacteria are (a) the ability to rapidly transport sugars, when in competition with other plaque bacteria, and (b) convert sugars rapidly to acid, even under extreme environmental conditions, such as at a low pH (Table 6.8). Few oral bacteria are able to tolerate acidic conditions for prolonged periods, but mutans streptococci and lactobacilli are not only able to remain viable (survive) at low pH, but are able to continue to metabolize and multiply, i.e. they are aciduric (acid-loving). That this combination of properties confers a selective advantage on these bacteria was shown in the laboratory using mixed culture competition studies (Table 6.9). At a constant pH 7, *S. mutans* and *L. casei* were non-competitive and were only minor components of the microbial community, even when exposed to pulses of fermentable sugar (glucose). However, when the pH was allowed to fall after a glucose pulse, *S. mutans* and *L. casei* gradually increased in proportions until they eventually dominated the mixed culture. As their proportions rose, so the rate and extent of the pH-fall increased. This rise of *S. mutans* and *L. casei* was at the expense of acid-sensitive species including *S. gordonii*, *N. subflava* and Gram-negative anaerobic rods.

Studies of fresh isolates of mutans streptococci have shown that strains of *S. sobrinus* are more acidogenic and more cariogenic in rodents than strains of *S. mutans*. However, this difference was lost when strains were repeatedly sub-cultured in the laboratory. No significant differ-

Table 6.7 Characteristics of mutans streptococci that contribute to its cariogenicity

Property	Comment
Sugar transport	High and low affinity transport systems operate over a wide range of conditions to ensure substrate uptake, even under extreme conditions, e.g. low pH
Acid production	An efficient glycolytic pathway rapidly produces low terminal pH values in plaque
Aciduricity	Cells are able to survive, metabolize and grow at low pH values
Extracellular polysaccharide (EPS) production	EPS contributes to the plaque matrix, consolidates attachment of cells, and may localize acidic fermentation products
Intracellular polysaccharide (IPS) production	IPS utilization allows acid production to continue in the absence of dietary sugars

Table 6.8 Acid production from glucose by oral bacteria

Bacterium	Terminal pH*	Rate† of acid production at	
		pH 7.0	pH 5.0
S. mutans	4.01	280	75
S. mutans (fresh isolate)	–‡	850	180
S. sobrinus	4.13	149	–
S. sobrinus (fresh isolate)	–	1700	400
S. gordonii	4.36	63	15
S. sanguis (fresh isolate)	–	900	0
S. mitis	4.34	167	–
A. naeslundii	4.57	57	–
A. viscosus	4.44	76	–
L. casei	3.86	164	–

* Terminal pH after 15 minutes
† Rates are expressed as nmol of acid produced per minute per mg (dry weight) of cells at a constant pH 7.0 or pH 5.0
‡ Not determined.

Table 6.9 Properties of bacteria in a mixed culture model system following 10 consecutive daily pulses of glucose, at a constant pH 7, or without pH control

Bacterium	Percentage viable count		
	Before glucose pulsing (pH 7)	After 10 glucose pulses	
		at constant pH 7	without pH control
S. mutans	0.3	1	19
S. gordonii	28	25	0.2
S. oralis	15	17	1
A. viscosus	0.1	13	2
L. casei	0.1	0.2	36
N. subflava	0.1	<0.1	0
V. dispar	10	29	41
P. intermedia	31	6	<0.01
F. nucleatum	15	10	<0.01
Terminal pH	7.0	7.0	3.8

ences have been found between mutans streptococci isolated from car-ies-active and caries-free humans. Following gnotobiotic animal studies, it was proposed that the presence of lactate-consuming bacteria in plaque might decrease the cariogenicity of mutans streptococci. More recent *in vitro* evidence has questioned this hypothesis (Chapter 5), while recent clinical studies have found *Veillonella* species in close association with mutans streptococci and lactobacilli during enamel demineralization, casting further doubt on their protective role.

Collectively, these findings allow a model to be constructed to explain the changes in the ecology of dental plaque that lead to the development of a caries lesion. Cariogenic bacteria may be found naturally in dental plaque, but at neutral pH, these organisms are weakly competitive and are present only as a small proportion of the total plaque community. In this situation, with a conventional diet, the processes of de- and re-mineralization are in equilibrium. However, if the frequency of ferment-able carbohydrate intake increases, then plaque spends more time below the critical pH for enamel demineralization (approximately pH 5.5; Figure 2.4). The effect of this is two-fold. Conditions of low pH favour the proliferation of mutans streptococci and lactobacilli, while tipping the balance towards demineralization. Greater numbers of mutans streptococci and lactobacilli in plaque would produce more acid at even faster rates, thereby enhancing demineralization still further. The rise in lactic acid production would also select for the growth of *Veillonella* spp. Other bacteria could also make acid under similar conditions, but at a slower rate. If these aciduric species were not present initially, then the repeated conditions of low pH coupled with the inhibition of competing organisms might increase the likelihood of colonization by mutans streptococci or lactobacilli. This sequence of events would account for the lack of total specificity in the microbial aetiology of caries and explain the pattern of bacterial succession observed in many clinical studies. This theory might be termed the **ecological plaque hypothesis**.

Summary

Overwhelming clinical evidence has found a positive relationship between mutans streptococci in plaque and the development of caries. This relationship is strongest for fissure caries, but is not unique. Early studies implicated *Actinomyces* species in the aetiology of root surface caries; more recent work has implicated mutans streptococci (and poss-ibly lactobacilli) with initial lesions. Evidence has also emerged that microbial succession may occur during the different stages of lesion development, both on enamel and on root surfaces. Mutans streptococci are associated with early demineralization while lactobacilli are implicated more with lesion progression and cavitation. The properties that confer pathogenicity to cariogenic bacteria are related to the ability to rapidly catabolize dietary carbohydrates to acid, and for the organisms to survive and proliferate under the fluctuating conditions of pH in plaque.

APPROACHES FOR CONTROLLING DENTAL CARIES

Mechanical removal of plaque by efficient oral hygiene procedures can almost completely prevent caries. Such measures are particularly effective when combined with a reduction in the amount and frequency of sugar intake. However, it is difficult to alter established eating habits and to maintain a high degree of motivation for effective oral hygiene. Alternative preventive measures are under development that require little co-operation from the public.

Fissure sealants

Occlusal pits and fissures are the most caries-prone areas of the human tooth. These areas are least affected by oral hygiene and anti-caries agents. Strongly adherent, self-polymerizing and UV-light polymerizing plastic sealant materials have been applied to fissures as a barrier against microbial attack. Problems surround the prolonged retention of these materials in the oral cavity and care must be taken to avoid sealing early caries lesions.

Fluoride

It has been known for over 40 years that the administration of fluoride systemically, via the water supply, can reduce the incidence of caries by at least 50%. The optimum concentration for maximal protection against caries is thought to be 1 part per million (1 ppm) but in many water supplies it naturally occurs at concentrations greater than this. Despite its proven value in decreasing caries incidence, the addition of fluoride to the drinking water remains a controversial and emotive issue. Fluoride can also be added to table salt, milk and toothpastes. It is also used in a variety of mouthwashes and gels for topical use and in tablets for supplementation of the systemic effect. In recent experiments, fluoride has also been incorporated into thin films for coating teeth, topical varnishes, and slow-release capsules. Fluoride is also available naturally in tea and the bones of fish (especially soft-boned sardines and salmon).

Ingested fluoride exerts its effect topically (if only transiently) and systemically after ingestion. Over 90% of available fluoride in the gut is absorbed into the blood stream, but well over 50% of this is rapidly excreted by the kidneys. The remaining absorbed fluoride is combined predominantly into skeletal tissue and unerupted and developing enamel with trace amounts being secreted into oral fluids (saliva, GCF). In oral fluids, fluoride can interact with the surface of the enamel of erupted teeth. The surface of the enamel is converted into crystalline forms of apatite containing high levels of fluoride, often referred to as fluorapatite. Fluorapatite is thermodynamically more stable than apatite and resists acid dissolution to a greater extent than hydroxyapatite. Thus, fluoride can protect both erupted and un-erupted teeth.

As was discussed in detail in Chapter 4, fluoride can also inhibit the metabolism of plaque bacterial populations. It achieves this by (1) reducing glycolysis by inhibition of enolase; (2) indirectly inhibiting sugar transport by blocking the PEP-PTS system; (3) acidifying the interior of cells, and hence inactivating key metabolic enzymes; (4) interfering with bacterial membrane permeability to ionic transfer and (5) inhibiting the synthesis of intracellular storage (IPS) compounds, especially glycogen.

Dental plaque has been found to concentrate fluoride from ingested water. In areas where the concentration of fluoride in the water supply is less than 0.01 ppm, dental plaque has been found to contain 5 to 10 ppm. In water supplies supplemented with 1 ppm fluoride, concentrations of up to 190 ppm have been found in dental plaque. Much of this fluoride is bound to organic components in plaque, but there is also evidence that it can be released when the pH falls, and be bioavailable to interfere with acid production by plaque bacteria. The sensitivity of oral bacteria to fluoride increases as the pH falls (Table 6.10), so that concentrations of fluoride that would be ineffective at resting pH values can be inhibitory at pH 5.0 or below. Mutans streptococci are particularly sensitive to low levels of fluoride at a moderately low environmental pH. Although surveys have failed to detect major changes in the qualitative and quantitative composition of plaque in humans residing in places with high or low natural levels of fluoride in the drinking water, fluoride is more likely to function prophylactically in these circumstances. Thus, mutans streptococci would be suppressed in plaque under conditions when they would otherwise be expected to flourish; the rate of change of pH in plaque would also be diminished. Laboratory evidence supports this hypothesis. Using the aforementioned mixed culture system, S. mutans remained a minor component of plaque (<1% of the cultivable microflora) during glucose pulsing in the presence of 1 mmol/l sodium fluoride when, under similar circumstances, it had previously attained levels of 19% in the absence of fluoride (Table 6.9). Fluoride can thus serve to stabilize the composition of the plaque microflora. In contrast, to the effects of natural levels of fluoride, artificially-high concentrations delivered locally by fluoride gels, for example, as used in the treatment of xerostomic patients, did lead to the elimination of mutans streptococci from the plaque of 10/33 patients after 5 years. Interestingly, lactobacilli are innately highly resistant to fluoride (Table 6.10) and were not affected by this gel therapy. In the laboratory, mutans streptococci can adapt to and tolerate even relatively high levels of fluoride. Concern has been expressed as to whether organisms might become resistant to fluoride due to the prolonged use of fluoride-containing toothpastes. However, laboratory studies have shown that fluoride-resistance reduces the acidogenicity and cariogenicity of mutans streptococci, so that even if adaptation occurred, there would be no increase in caries risk.

Table 6.10 The minimum inhibitory concentration (MIC) of oral bacteria to sodium fluoride at neutral and low pH

Bacterium	MIC (mmol/l)	
	pH 7.0	pH 5.0
S. mutans	7.5	1.0
S. gordonii	7.5	1.0
S. oralis	7.5	2.5
A. viscosus	10.0	0.5
L. casei	>10.0	10.0
N. subflava	>10.0	NG*
V. dispar	>10.0	2.5
P. intermedia	5.0	0.1
F. nucleatum	5.0	NG*

* N. subflava and F. nucleatum would not grow at pH 5.0, even in the absence of sodium fluoride.

There has been a dramatic fall in the prevalence of caries in many industrialized societies since the mid–1970s. This corresponds to the introduction of fluoride-containing toothpastes, and much of the benefit derived from fluoride is attributable to its delivery from this source. Although much of the anti-caries benefit of fluoride is due to topical and systemic effects on enamel, it is likely that there is also an antimicrobial effect. This antimicrobial effect can be enhanced by changing the counter-ion; thus, stannous fluoride is markedly more inhibitory to oral bacteria than sodium fluoride. Metal salts can be unstable in tooth-paste formulations, but recently several products have been launched containing stannous fluoride. Other antimicrobial agents will be dis-cussed in the next section.

Antimicrobial agents

The use of antimicrobial agents to control plaque has been advocated for a number of years. Many compounds have been delivered via mouthrinses but a number of products are now being formulated successfully into toothpastes. The aim of these products is to be anti-plaque (and hence also anti-gingivitis), rather than just anti-caries, but a number of issues relevant to the control of caries will be discussed here.

Although dental caries can be prevented by antibiotics, these inhibi-tors are considered inappropriate for routine, unsupervised use in toothpastes or mouthrinses. Their spectrum of activity is too broad, while their indiscriminate use can lead to the emergence of resistant organisms, and to overgrowth by opportunistic pathogens (such as C. albicans) due to the suppression of the resident microflora. Mouthrinses have proved to be a successful vehicle for the delivery of anti-plaque agents. Usually, the mouthrinse is a water/alcohol mixture to which flavourings, a non-ionic surfactant, and a humectant, together with an antimicrobial agent, are added. These agents include quaternary ammonium compounds, bisbiguanides, enzymes, metal salts, 'essential oils' and plant extracts; some examples are listed in Table 6.11. Mouthrinses have efficacy against salivary bacteria and supragingival plaque only; because of the relatively short contact time between the inhibitor in the rinse and the mouth, the agents must bind to oral surfaces. The property is termed substantivity and involves electrostatic (ionic and covalent binding), hydrogen bonding, and van der Waals (lipophilic binding) forces. Once adsorbed, effective inhibitors are released slowly into the oral environment (especially saliva) and inhibit the growth or metabolism of micro-organisms for prolonged periods even at sub-MIC concentrations.

The most effective agent to date has been chlorhexidine which has a broad spectrum of activity against yeasts, fungi, and a wide range of Gram-positive and Gram-negative bacteria. Chlorhexidine can reduce plaque, caries, and gingivitis in humans; it is not used for prolonged periods because side-effects such as staining of teeth can occur. At high concentrations, chlorhexidine is bactericidal and acts as a detergent by damaging the cell membrane. The benefits of chlorhexidine are not limited to this initial bactericidal effect, as other agents with a similar MIC do not have the same anti-plaque properties. Chlorhexidine is

Table 6.11 Some classes and examples of inhibitors used as anti-plaque agents in mouthrinses and toothpastes

Class of inhibitor	Examples
Bisbiguanide	chlorhexidine, alexidine
Enzymes	mutanase, glucanase; amyloglucosidase-glucose oxidase
'Essential oils'	thymol, eucalyptol
Metal ions	copper, zinc, stannous
Plant extracts	sanguinarine
Phenols	triclosan
Quaternary ammonium compounds	cetylpyridinium chloride
Surfactants	sodium lauryl sulphate

substantive, and hence is bound to oral surfaces from where it is released gradually into saliva over many hours at bacteriostatic concentrations. At these concentrations, chlorhexidine can reduce acid production in plaque. Laboratory studies have shown that these sub-MIC concentrations can (1) abolish the activity of the PEP-PTS sugar transport system (Chapter 4) and thereby markedly inhibit acid production in streptococci; (2) inhibit amino acid uptake and catabolism in *S. sanguis*; (3) inhibit the trypsin-like protease of *P. gingivalis*, and (4) affect various membrane functions, including the ATP synthase and the maintenance of ion gradients in streptococci. Chlorhexidine can also affect microbial pathogenicity in animal models when used at sub-MIC levels.

Mutans streptococci are more sensitive to chlorhexidine than other oral streptococci, and this property has been exploited in those people with a high risk of developing caries. Surveys in Scandinavia showed that subjects with levels of mutans streptococci that were $>10^6$ CFU/ml saliva had an increased risk of developing caries. Salivary concentrations of mutans streptococci could be reduced by professional oral hygiene, dietary counselling, topical fluoride, and also by the use of chlorhexidine mouthrinses. While mutans streptococci were suppressed, other oral streptococci such as *S. sanguis* (which is associated more with oral health) were relatively unaffected. This approach has also been applied successfully to expectant mothers. The suppression of mutans streptococci in mothers reduced the transmission of these potentially cariogenic organisms to the baby and delayed the onset of caries. Chlorhexidine has also been applied to teeth as a varnish. These varnishes have led to the successful elimination of mutans streptococci for prolonged periods. Other antimicrobial agents used in mouthrinses include cetylpyridinium chloride, sanguinarine, thymol, menthol and sodium lauryl sulphate.

Toothpastes (dentifrices) are another vehicle for the unsupervised delivery of antimicrobial agents. However, many proven antimicrobial and antiplaque agents, such as chlorhexidine, are incompatible with components of toothpastes, and lose their bio-activity. A new phase of toothpaste development is underway based on the addition of antimicrobial agents, either alone or in combination, to provide clinical benefit above that obtained just from fluoride and detergents. Some of these toothpastes are also formulated to give an anti-gingivitis effect, and several products include Triclosan. Triclosan also has a broad spectrum of antimicrobial activity against yeasts and Gram-positive and Gram-negative bacteria. Like chlorhexidine, it is also substantive and multi-functional in its mode of action. At sub-MIC concentrations, it can inhibit acid production by streptococci and protease activity by *P. gingivalis*. The anti-plaque activity of Triclosan has been enhanced by combining it with (1) a copolymer to boost its oral retention, or (2) with zinc citrate. Zinc ions are also substantive and can inhibit sugar transport, acid production, and protease activity. Combinations of agents (e.g. zinc citrate and Triclosan) can give additive clinical benefit. Enzymes have also been included in toothpastes, although they pose considerable formulation difficulties to prevent their precipitation and inactivation by detergents. Dextranases and glucanases isolated from fungi have been used to modify the plaque matrix and reduce plaque formation, while glucose oxidase and amyloglucosidase have been used to boost the activity of the salivary peroxidase (sialoperoxidase) system (Chapter 2). A variety of plant extracts have potentially valuable properties; sanguinarine and tannins have antiplaque activity and can inhibit glycolysis and glucosyltransferase activity, respectively.

A number of products are formulated to reduce calculus formation (Chapter 5); inhibitors of mineralization include polyphosphonates, zinc salts, and pyrophosphates. Future developments might see the application of surface active agents to prevent or reduce bacterial colonization of teeth, or the use of monoclonal antibodies targeted against specific organisms. Substituted aminoalcohols have been found to interfere with bacterial attachment and to partially disperse pre-formed dental plaque; their delivery from toothpastes is currently under investigation. Passive immunization with monoclonal antibodies against *S. mutans* was not only successful in eliminating these organisms from plaque, but also prevented their recolonization for periods of over a year.

A consequence of the regular, unsupervised use of antimicrobial agents from toothpastes and mouthrinses is that their prolonged use should not lead to the disruption of the ecology of the oral microflora by either (a) perturbing the balance among the resident organisms, which might lead to the overgrowth by potentially more pathogenic species; or (b) the development of resistance. Guidelines are now laid down to ensure that manufacturers perform long-term clinical trials to confirm that these eventualities do not occur.

Sugar substitutes

Most humans enjoy and prefer to eat sweet substances. Unfortunately many sweet foods are composed of mono- or disaccharides which are easily metabolized by plaque bacteria and predispose enamel to dental caries. To satisfy the human preference for sweet substances without causing caries the use of inert (non-metabolizable) dietary sweeteners has been proposed. Artificial sweeteners are of two types, the intense type many times sweeter than sucrose, and the bulk agents which are usually not as sweet. The intense sweeteners include cyclamate, aspartame and saccharin. These agents are extremely sweet and their use has been proposed for drinks. The bulk agents, e.g. mannitol and sorbitol, are not as sweet as sucrose and are therefore used in the confectionery industry. The use of the bulk agents mannitol and sorbitol is based on the premise that they cannot be metabolized by the majority of plaque bacteria. However, some oral bacteria, e.g. mutans strepto-cocci, can metabolize these two sugar alcohols to acid, albeit slowly (Figure 6.3). The frequent use of sorbitol-containing products can also lead to increases in the numbers of sorbitol-fermenting bacteria, in general, and in mutans streptococci, in particular. The mean pH fall in plaque from sorbitol or glucose also increased following regular exposure to sorbitol. Xylitol is another bulk sweetener, but it is not metabolized by plaque bacteria (Figure 6.3). In human trials, xylitol can reduce significantly the incidence of caries, both by reducing the frequency of acid attack on the enamel (Figure 2.4) and by stimulating saliva flow, thereby encouraging remineralization. There may be another mechanism by which xylitol reduces the caries potential of plaque. Xylitol is transported into cells of $S.$ $sobrinus$ by the fructose-PTS where it enters a 'futile cycle' of phosphorylation, dephosphoryl-ation and expulsion. This futile cycle reduces the rate of growth and acid production (from exogenous sugars such as glucose) of cells, and leads to reduced levels of mutans streptococci and caries in habitual users of xylitol-containing confectionery. Xylitol interferes with sugar metabolism of mutans streptococci by consuming PEP and NAD^+ during the 'futile cycle' and competitively inhibiting glycolysis at the phosphofructokinase level by either xylitol–5-phosphate or xylulose–5-

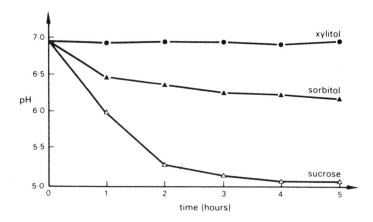

Figure 6.3 Typical fall in pH ('Stephan curve') of plaque suspended in buffer incubated in the presence of 0.75% (w/v) xylitol, sorbitol and sucrose.

phosphate. Lycasin, a hydrogenated corn-starch syrup, has also been incorporated into confectionery with some success. Like xylitol, it stimulates saliva flow and induces enamel remineralization; it has also been shown to be capable of reversing the development of white-spot lesions in humans.

Vaccination

The oral cavity is provided with all of the components necessary to mount an effective immune response against micro-organisms (Chapter 2). While the microbial aetiology of dental caries is not totally specific, considerable evidence implicates mutans streptococci as the major group of causative bacteria. These facts have led to the concept of vaccinating against dental caries using mutans streptococci as an immunogen. Indeed, studies using a vaccine based on *S. mutans* were initiated in Great Britain over 20 years ago. The early studies used crude whole cell preparations of *S. mutans* to protect monkeys. Although protection against caries and a reduction in the levels of mutans streptococci was achieved, concern was expressed over possible immunologically-mediated tissue damage in humans following exposure to streptococcal antigens (as occurs in, for example, rheumatic fever). Indeed, in some animals vaccinated with whole cells of *S. mutans*, antibodies were formed which reacted not only with the bacteria, but also with heart tissue. The significance for humans was difficult to assess, but it was clear that such a cross-reaction was undesirable, and that there existed a risk of damage to the heart. Subsequent work has been directed towards characterizing the antigenic composition of mutans streptococci and selecting individual purified antigens that will confer protection but lack the human tissue cross-reactivity.

An alternative animal model was adopted in the USA to study anti-dental caries vaccines. Several groups used the development of caries on smooth surfaces by *S. mutans* serotype *d* (now classified as *S. sobrinus*; Chapter 3) in rats as a model system. Rodents have the advantage that unlike monkeys, they are cheap and easy to maintain so that large experimental groups can be compared. However, there are major differences in the morphology of teeth and composition of saliva that may make rodents a less appropriate model of human caries than monkeys. Also, the development of caries by *S. sobrinus* is more by sucrose-mediated mechanisms, so that sub-unit vaccines in this model are directed towards generating antibodies against glucosyltransferases (GTF). Immunization of rats or hamsters with GTF preparations led to reduced plaque accumulation and caries, particularly on smooth surfaces. However, such vaccines did not give a good protective effect in monkeys, raising doubts about their potential effectiveness in humans.

Two purified proteins present on the cell surface of *S. mutans* (serotype *c*) have been developed from studies of the monkey model as immunogens for human anti-caries vaccines. Antigen I/II (also referred to as antigen B, or P1) is a high molecular weight (M. Wt. = 185 000) protein that has been shown to confer protection in monkeys. It is found in *S. mutans*, and other species of oral streptococci, and so could give wide ranging protection. Any doubts as to whether this molecule could also give rise to tissue cross-reacting antibodies have been dispelled by the generation of peptides from the native antigen and assaying for the smallest fragment that confers protection without side effects. A peptide with a molecular weight of only 3 800 has been found to confer greater protection when topically applied to the gingivae than the parent molecule. The antigen stimulated IgG antibodies in GCF and sIgA antibodies in saliva. It has been proposed that the protective mode of action might involve preventing the adherence of *S. mutans* to teeth. Synthetic peptides have been prepared that retain the protective activity. As discussed earlier, passive immunization could be achieved in monkeys and humans using monoclonal antibodies against antigen I/II. The antibodies prevented colonization of teeth by either exogenous or indigenous *S. mutans*.

Another vaccine candidate is antigen A (M.Wt. = 29 000), which has been purified using affinity chromatography on monoclonal antibodies, and has been shown to be protective in monkeys. This antigen has never been associated with cross-reactivity with human tissue. Recent studies have also raised the possibility of oral immunization using water-soluble and insoluble GTF from *S. sobrinus*. Preliminary studies using human volunteers have shown sIgA anti-GTF antibodies were induced following the administration of GTF in capsules; these antibodies also appeared to reduce the accumulation of indigenous mutans streptococci. Other surface antigens of mutans streptococci that have been considered as contenders for inclusion in a vaccine include FTF, glucan-binding protein, and antigens C and D (other cell surface proteins).

The optimum route by which any vaccine might be administered for use in human trials has yet to be resolved. The possibilities include systemic injection (to generate circulating antibodies, especially of the IgG class, which would enter the mouth via GCF), oral vaccines, and local immunization (which elicit mainly an sIgA response), and passive immunization. The host response will be influenced by the frequency and level of dosage needed, the age of the subject, and perhaps their prior experience of exposure to *S. mutans* antigens.

Although potential vaccines against *S. mutans* have been manufactured to standards that satisfy the legislative authorities, there have been no major field trials to assess their efficacy in humans. The main reasons for this are two-fold. During the time of the development of these vaccines, the incidence of caries has fallen dramatically in most industrialized societies, probably as a result of fluoride. Therefore, the need for a vaccine in these countries has diminished. Also, the public acceptance of new mass-vaccination programmes can be poor, even for serious medical infections. A major question facing health organizations is whether a vaccine is justified against a non-life threatening disease? It may be that vaccination could be considered of benefit to particular high-risk groups in the population. Despite this hold-up, research continues to explore the mechanisms behind protection and to develop even more refined sub-unit vaccines.

SUMMARY

Numerous cross-sectional and longitudinal surveys have found a strong association between dental caries and the levels of mutans streptococci in dental plaque. This association is strongest for fissure and rampant caries; the evidence for approximal surfaces is less strong, possibly due to the problems of sampling plaque and diagnosing lesions at this inaccessible site. Early studies found an association between *Actinomyces* spp. and root surface caries; more recent work has reported stronger correlations with mutans streptococci and lactobacilli. The development of lesions on all surfaces appears to involve different waves of bacterial succession. Mutans streptococci are implicated in caries initiation while lactobacilli and possibly *A. odontolyticus* appear to be related to lesion progression. The association of mutans streptococci with dental caries is not unique; in some cases, caries can develop in their apparent absence, and they can persist at sites without demineralization. The progression of lesions can result in the invasion of the dentine and pulp. When infected, a diverse collection of organisms can be isolated from these sites. The predominant organisms are often Gram-positive rods, although proteolytic Gram-negative species have also been recovered.

The properties of cariogenic bacteria that appear to correlate with their pathogenicity are the ability to rapidly metabolize dietary carbohydrates to acid over a range of environmental conditions, but especially at low pH, and to be able to survive and grow under the acidic conditions so generated (i.e. the combined properties of acidogenicity and aciduricity). Additional properties that may also play a role include the synthesis of intra- and extracellular polysaccharides. Strategies to control or prevent dental caries are based on (a) reducing levels of either plaque in general, or specific cariogenic organisms in particular, by antiplaque and antimicrobial agents; (b) using fluoride to strengthen the resistance of enamel to acid attack and (c) inhibiting acid production by avoiding the frequent intake of fermentable carbohydrates, by replacing such carbohydrates with alternative sweetening agents (sugar substitutes), or by interfering with plaque metabolism with fluoride or antimicrobial agents. Immunization against mutans streptococci using sub-unit vaccines is also a potential route for protection. Despite the fall in incidence of dental caries over the past two decades, enamel caries remains a highly prevalent disease in young people, while root-surface caries and recurrent caries are an increasing problem for older individuals.

FURTHER READING

Adriaens, P. A., Loesche, W. J. and De Boever, J. A. (1986) Bacteriological study of the microbial flora invading the radicular dentine of periodontally diseased caries-free human teeth, in *Borderland between Caries and Periodontal Disease III*. (eds. T. Lehner and G. Cimasoni) Editions Médecine et Hygiène, Geneva, pp. 383–90.

Bowden, G. H. W. (1990) Microbiology of root surface caries in humans. *Journal of Dental Research*, **69**, 1205–10.

Boyar, R. M. and Bowden, G. H. (1985) The microflora associated with the progression of incipient lesions in teeth of children living in a water fluoridated area. *Caries Research*, **19**, 298–306.

Dawes, C. and Ten Cate, J. M. (eds.) (1990) Fluorides: Mechanisms of action and recommendations for use. *Journal of Dental Research*, **69**, 505–835.

Gjermo, P. (1989) Chlorhexidine and related compounds. *Journal of Dental Research*, **68**, 1602–8.

Guggenheim, B. (ed) (1979) *Health and Sugar Substitutes*. Karger, Basel.

Hamilton, I. R. and Bowden, G. H. (1988) Effect of fluoride on oral micro-organisms, in *Fluoride in Dentistry*. (eds. J. Ekstrand, O. Fejerskov and L. M. Silverstone) Munksgaard, Copenhagen, pp. 77–103.

Hardie, J. M., Thomson, P. T., South, R. J., Marsh, P. D., Bowden, G. H., McKee, A. S., Fillery, E. D. and Slack, G. L. (1977) A longitudinal epidemiological study of dental plaque and the development of dental caries – interim results after 2 years. *Journal of Dental Research*, **56**, C90–8.

Huis int' Veld, J., Van Palenstein-Helderman, W. H. and Dirks, O. B. (1979) *Streptococcus mutans* and dental decay in humans – a bacteriological and immunological study. *Antonie van Leeuwenhoek*, **45**, 25–33.

Krasse, B., Emilson, C. and Gahnberg, L. (1987) An anticaries vaccine: report on the status of research. *Caries Research*, **21**, 255–76.

Loesche, W. J. (1986) Role of *Streptococcus mutans* in human dental decay. *Microbiological Reviews*, **50**, 353–80.

Loesche, W. J. and Straffon, L. H. (1979) Longitudinal investigations of the role of *Streptococcus mutans* in human fissure decay. *Infection and Immunity*, **26**, 498–507.

Marsh, P. D., Featherstone, A., McKee, A. S., Hallsworth, A. S., Robinson, C., Weatherell, J. A., Newman, H. N. and Pitter, A. F. V. (1989) A microbiological study of early caries of approximal surfaces in schoolchildren. *Journal of Dental Research*, **68**, 1151–4.

Russell, R. R. B. and Johnson, N. W. (1987) The prospects for vaccination against dental caries. *British Dental Journal*, **162**, 29–34.

Scheie, A. AA. (1989) Modes of action of currently known chemical anti-plaque agents other than chlorhexidine. *Journal of Dental Research*, **68**, 1609–16.

Silverstone, L. M., Johnson, N. W., Hardie, J. M. and Williams, R. A. D. (1981) *Dental Caries: Aetiology, Pathology and Prevention*. Macmillan, London.

Van Houte, J., Gibbs, G. and Butera, C. (1982) Oral flora of children with 'nursing bottle caries'. *Journal of Dental Research*, **61**, 382–5.

7 Periodontal diseases

The term 'periodontal diseases' embraces a number of conditions in which the supporting tissues of the teeth are attacked. Although it is more than 300 years since Leeuwenhoek implicated micro-organisms with oral malodours and inflammation of the gums, it is only in the past two decades that any real progress has been made in identifying and characterizing the bacteria that are associated with the periodontium in health and disease. The classification of periodontal diseases is still far from being resolved, but one terminology relates to the age group of the person affected (e.g. pre-pubertal, juvenile, adult), the rate of progress of the disease (rapid, acute, chronic), the distribution of lesions (localized or generalized), or whether there are any particular debilitating or predisposing factors (e.g. pregnancy, diabetes, HIV-infection).

In periodontal diseases, the junctional epithelial tissue at the base of the gingival crevice migrates down the root of the tooth to form a periodontal pocket (Figures 2.1 and 7.1). The predominant micro-organisms in disease are frequently obligately anaerobic or carbon dioxide-requiring (capnophilic) Gram-negative rods, filaments, or spiral-shaped bacteria, many of which are nutritionally-fastidious and difficult to grow in the laboratory. Consequently, special precautions have to be taken when sampling and processing the microflora from periodontal pockets to ensure that the viability of all of the representatives of the microflora is maintained.

ECOLOGY OF THE PERIODONTAL POCKET: IMPLICATIONS FOR PLAQUE SAMPLING

As discussed in Chapters 2, 4 and 5, the ecology of the gingival crevice is different to that of other sites in the mouth; it is more anaerobic and the site is bathed in gingival crevicular fluid (GCF). In disease, the crevice becomes a pocket, and the Eh has been shown to fall even lower, i.e. the site becomes even more anaerobic, while the flow of

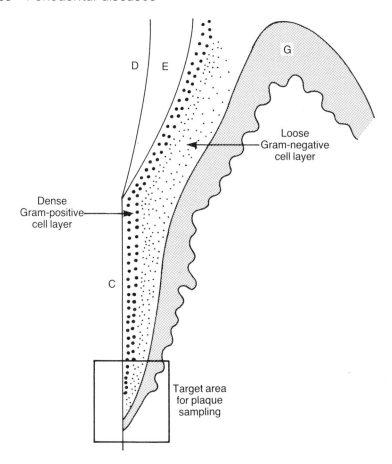

Loose
Gram-negative
cell layer

Dense
Gram-positive
cell layer

Target area
for plaque
sampling

Figure 7.1 Anatomy of the periodontal pocket. A diagrammatic representation of a periodontal pocket illustrating the difficulty in sampling the advancing front of the lesion without removing bacteria from other sites in the pocket. Key: G, gingiva; E, enamel; D, dentine; C, cementum.

GCF is increased. GCF will influence the pocket microflora in a number of ways. It contains most of the humoral and cellular defence factors found in serum, and also has a number of complex proteins and glycoproteins that can serve as novel substrates for bacterial metabolism (Table 7.1). These include iron and haeme-containing molecules and, as discussed in Chapter 4, pocket bacteria interact synergistically to degrade these substrates. Other bacterial growth factors that can be present include urea, α–2-globulin (which can be used by *T. denticola*), and in females, hormones such as progesterone and oestradiol (which are used by black-pigmented anaerobes). Unlike dental caries, many of the bacteria associated with periodontal diseases are asaccharolytic but proteolytic. A consequence of proteolysis is that the pH in the pocket during disease becomes slightly alkaline (pH 7.4–7.8) compared to health (pH 6.9). The growth and enzyme activity of some periodonto-pathogens, such as *Porphyromonas gingivalis*, is probably enhanced by such conditions.

Table 7.1 Components of Gingival Crevicular Fluid (GCF) that might affect the composition of the microflora of the periodontal pocket

Host defences	Novel nutrients
IgG	Haemin; iron
IgA	Albumin
IgM	α–2-globulin
	Transferrin
Complement	Haemopexin
	Hormones
B and T lymphocytes	Haptoglobin
Neutrophils	Haemoglobin
Macrophages	Proteins; glycoproteins

The flow of GCF can remove micro-organisms not attached firmly to a surface. The cementum surface of the tooth is colonized by Gram-positive bacteria belonging to the genera *Streptococcus* and *Actinomyces*. Many putative periodontopathogens (*Prevotella, Porphyromonas, Fusobac-terium* spp.) attach to this layer of cells by coaggregation (Chapter 4). Likewise, some of the black-pimented anaerobes, including *Porphyro-monas gingivalis, Prevotella melaninogenica* and *P. intermedia*, as well as *Peptostreptococcus micros* may persist in the pocket due to their ability to adhere to crevicular epithelial cells. Indeed, their attachment to these cells is markedly enhanced when the epithelium has been treated with proteases. Thus, proteolytic enzymes of either bacterial or host origin might promote the colonization of these putative periodontopathogens.

When attempting to determine the microflora of a periodontal pocket, care has to be taken to preserve the viability of the obligately anaerobic species during the taking, dispersing, diluting and cultivation of the sample (Chapter 4). The sample should be taken from the base of the pocket, near the advancing front of the lesion in order to avoid remov-ing organisms that are not associated with tissue destruction (Figure 7.1). Methods include the insertion of paper points (although this approach will not remove the strongly adherent bacteria), currettes or scalers, and barbed broaches held in a protective cannula and flushed continuously with oxygen-free gas. Samples are usually delivered to the laboratory in specialized transport fluids, designed to maintain a low redox potential while minimizing cell division, and then processed in an anaerobic chamber. The use of such a rigorous approach has led to the recovery and discovery of organisms never before described. This has been one of the reasons for the massive changes in the taxonomy of oral micro-organisms over recent years (Chapter 3). Valid classification of organisms is essential if the association of species with particular diseases is to be made. Results of some of the studies on the microbial aetiology of periodontal diseases will now be presented, but first the evidence for the role of microbial involvement in these diseases will be discussed.

EVIDENCE FOR MICROBIAL INVOLVEMENT IN PERIODONTAL DISEASES

Evidence that bacteria are implicated in periodontal diseases has come from gnotobiotic animal studies. Germ-free animals rarely suffer from periodontal disease, although, on occasions, food can be impacted in the gingival crevice producing inflammation. However, inflammation is much more common and severe when specific bacteria, particularly some of those isolated from human periodontal pockets, are used in pure culture to infect the animals. These bacteria include streptococci and *Actinomyces* spp. but are more commonly Gram-negative, for example, *Actinobacillus, Prevotella, Porphyromonas, Capnocytophaga, Eikenella, Fusobacterium* and *Selenomonas* spp. Furthermore, periodontal disease is arrested when an antibiotic active against the particular organism is administered to the infected animal. In humans, evidence for the role of micro-organisms has also come from plaque control and antibiotic treatment studies. However, these latter types of studies give no information as to whether disease results from the activity of (a) a single, or only a limited selection, of species (the specific plaque hypothesis); or (b) any combination of a wide range of plaque bacteria (the non-specific plaque hypothesis). In order to test these 'hypotheses', a large number of cross-sectional epidemiological studies have been performed on patients with particular forms of periodontal disease. As with dental caries, a disadvantage of this type of study is that true 'cause-and-effect' relationships can never be determined. Organisms that appear to predominate at diseased sites might be present as a result of the disease, rather than having actually initiated it. With the exception of gingivitis, longitudinal studies (which do not suffer from this drawback) are not usually possible because of the lengthy natural history of some forms of periodontal disease and the difficulties in predicting subjects and sites likely to become affected. Despite these problems of study design, coupled with the technical difficulties mentioned earlier, genuine progress has been made in our understanding of the microbiology of human periodontal diseases, and the principal findings will now be presented.

MICROBIOLOGY OF PERIODONTAL DISEASES

Some of the different forms of periodontal diseases are listed in Table 7.2. The principal microbiological findings from the more common types of disease will be discussed first, followed by data from other, more rare conditions.

Table 7.2 Different types of periodontal diseases

Common	Rare
Chronic marginal gingivitis	HIV-gingivitis and periodontitis
Chronic adult periodontitis	Pregnancy gingivitis
	Acute streptococcal gingivitis
	Viral gingivitis
	Acute necrotizing ulcerative gingivitis
	Juvenile periodontitis
	Pre-pubertal periodontitis
	Rapidly progressing periodontitis

Chronic marginal gingivitis

Chronic marginal gingivitis is a non-specific inflammatory response to dental plaque involving the gingival margins. If good oral hygiene is restored, it is usually eradicated and the gingival tissue becomes clinically normal again (Figure 7.2). Estimates of its incidence are difficult to determine but probably the whole dentate population is affected by this condition at some stage. Generally, gingivitis is regarded as resulting from a non-specific proliferation of the normal gingival crevice microflora due to poor oral hygiene. Certainly, the number of bacterial cells in plaque associated with gingivitis is increased substantially by 10–20 fold compared with healthy control sites. However, this increase in plaque mass is associated with increases in the proportions of only

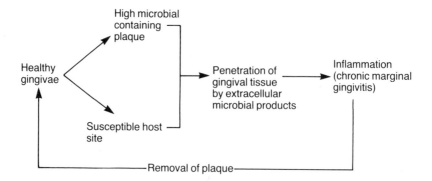

Figure 7.2 Schematic diagram of the aetiology of chronic marginal gingivitis.

a limited number of bacterial types which are mainly *Actinomyces* spp., plus certain facultatively and obligately anaerobic Gram-negative rods. The finding in the 1960s that gingivitis develops in a predictable and reproducible manner in volunteers who refrain from oral hygiene measures has allowed the design of longitudinal studies to determine the bacteriological events that lead to disease. The validity of the 'experimental gingivitis' model in humans has been confirmed in that no major differences have been found in the microflora of naturally-occurring gingivitis and that of human experimental gingivitis. However, some differences in microflora have been found between experimental gingivitis in children and adults, but not between natural gingivitis in the same age groups. The microflora associated with gingivitis is more diverse and differs in overall composition from that found in health. There is an increase in plaque mass, and there is a shift from the streptococci-dominated plaque of gingival health (Chapter 5) to one in which *Actinomyces* spp. predominate. The proportions of capnophilic (especially *Capnocytophaga* spp.) and obligately anaerobic Gram-negative bacteria also increase. An early study of 25 subjects suggested that specific relationships might exist between certain bacteria and particular stages in the development of gingivitis. When the gingivitis score was plotted as a function of the amount of plaque (plaque score), the gingivitis score increased in two large increments. For example, proportions of *A. israelii* increased significantly from 13 to 26% of the total cultivable microflora with the onset of a non-bleeding gingivitis. This was also associated with an increase in the percentage viable counts of *A. viscosus* from 7 to 14% of the microflora. When the gingivitis progressed to a bleeding stage, black-pigmented anaerobes increased from 0.01 to 0.2% of the microflora. The possible effect that bleeding might have on the sub-gingival microflora is significant because black-pigmented anaerobes require haemin for growth and this can be derived from the degradation of proteins and glycoproteins present in GCF (Table 7.1).

The potential diversity of the sub-gingival microflora, and the difficulties associated with data analysis, can be gauged from the results from two of the most detailed and comprehensive microbiological studies to date. Over 160 different bacterial groups (taxa) were cultivated from four young adults participating in an experimental gingivitis study, while more than 100 non-spirochaetal taxa were isolated from 21 children and adults with naturally-occurring gingivitis. In the former study, of the 166 taxa isolated, 73 showed a positive correlation with gingivitis, 29 were negatively correlated while the remainder either showed no correlation or were regarded as being present as a result of gingivitis. This last conclusion emphasizes the value of longitudinal studies. Despite the variability in the composition of the microflora between subjects, certain trends have emerged. The microflora becomes more diverse with time as gingivitis develops although none of the taxa from either study were uniquely associated with gingivitis. However, some organisms are found more commonly in gingivitis and are rare in health. Some of the most likely aetiological agents in the experimental gingivitis study in young adults are shown in Table 7.3. In addition, *P. intermedia*, curved rods (*Campylobacter* and *Wolinella* spp.), *S. sanguis*, *Peptostreptococcus anaerobius*, certain *Eubacterium* spp., and other spirochaetes were isolated in increased numbers in naturally-occurring gingivitis in adults and children. Interestingly, children with gingivitis had fewer sites with spirochaetes than adults with gingivitis, although the positive samples from children contained greater proportions of the spirochaete, *T. socranskii*, subsp. *paredis*. Increased proportions of *F. nucleatum* have also been reported in gingivitis; *A. actinomycetemcomitans* has also been found in low numbers but *P. gingivalis* is rarely isolated. It is still not clear whether gingivitis is a necessary stage for the development of more serious forms of periodontal disease or whether these can arise independently. Certainly some species that predominate in periodontitis, but which are not detectable in the healthy gingiva, were found as a small percentage of the microflora in gingivitis suggesting that environmental conditions which develop during gingivitis (e.g. bleeding, increased flow of GCF) may favour the growth of species implicated in periodontitis.

Chronic adult periodontitis

This is the most common form of advanced periodontal disease affecting the general population; it has also been described as chronic inflamma-

Table 7.3 Predominant bacteria of experimental gingivitis in young adult humans

Bacteria	Mean percentage viable count
Actinomyces israelii	1.4
Actinomyces naeslundii*	2.7–6.9
Actinomyces naeslundii/viscosus	4.0
Actinomyces odontolyticus*	1.7–3.7
Propionibacterium acnes	1.4
Lactobacillus sp.	2.1
Streptococcus anginosus	5.5
Streptococcus mitis	1.0
Peptostreptococcus micros	1.4
Anaerobic coccus	0.6
Prevotella oris	1.2
Veillonella parvula	11.7
Treponema sp.	+†
Fusobacterium nucleatum	4.4

* More than one biotype was recognized
† Present, but not enumerated.

tory periodontal disease. It differs from chronic marginal gingivitis in that in addition to the gingivae being involved, there is loss of attachment between the root surface and alveolar bone, and bone loss itself may occur. Gingivitis does not necessarily progress to periodontitis, and although periodontitis is usually preceded by gingivitis this is not always the case. Factors that enhance plaque retention or impede plaque removal, such as sub-gingival calculus, overhanging restorations or crowded teeth, predispose towards chronic periodontitis.

The inflammatory response to plaque is a basic host defence mechanism against microbial infections. Unfortunately this host response can also contribute to the destruction of tissues due to, for example, the release of lysosomal enzymes during phagocytosis or to the production of cytokines that stimulate resident connective tissue cells to release metalloproteinases. However, the effectiveness of the inflammatory response can be gauged by the overall slow rate of progression of the lesion compared to periodontal diseases in individuals with impaired host defences, so that tooth loss may take several decades to occur. During this period, there will be spells of disease activity followed by quiescent times, coupled with healing. Nevertheless, chronic periodontitis is common in humans and, without intervention, is a major cause of tooth loss after the age of 25 years.

Early studies of plaque associated with chronic periodontitis relied on microscopically-observed, qualitative morphological descriptions of the micro-organisms present. Dark-field microscopy showed that many of the bacteria in plaque from patients with deep pockets were motile and that spirochaetes could be present in high numbers. Some of these motile organisms have now been identified in cultural studies and shown to be *C. rectus* (formerly *W. recta*) and *S. sputigena*. Initially, it was believed that numbers of motile and spiral-shaped bacteria correlated with disease activity, and could form the basis of a cheap and rapid test for use in the clinic to monitor the status of a pocket. However, many putative periodontal pathogens cannot be distinguished on the basis of their cell morphology, and it is now less clear whether the increases in spirochaetes and motile organisms cause or merely follow disease progression. Consequently, microscopy found fewer applications in periodontal research until immunological probes against key periodontopathogens were developed. Now the presence of a limited number of selected bacteria can be screened for directly in plaque samples without the need for lengthy cultural procedures. These techniques are being used by some to monitor the effectiveness of treatment, or to determine which antibiotic to use in refractory cases.

The major approach to determine possible aetiological agents in chronic adult periodontitis has been to use conventional culture techniques to identify the predominant members of the plaque microflora. Numerous studies, predominantly cross-sectional, have been performed on different patient groups with pockets of varying depths (the deeper the pocket, the more severe or advanced the lesion) from a variety of geographical areas. All studies agree that the microflora is diverse and is composed of large numbers of obligately anaerobic Gram-negative rod and filament-shaped bacteria, many of which are asaccharolytic but proteolytic. These bacteria are often difficult to recover and identify in the laboratory and there is often conflicting evidence as to which organisms are the primary pathogens. A major difficulty in handling the large quantities of data from these studies was overcome when cluster analysis techniques were applied. This approach implicated certain clusters or combinations of bacteria with disease. Significantly, very different clusters of bacteria appeared to be able to produce an apparently similar pathological response. However, in advanced chronic periodontitis, clusters appeared to be dominated by either *P. gingivalis* or *F. nucleatum*. Thus, unlike some of the acute forms of periodontal disease, chronic periodontitis appears to result from the activity of *mixtures* of interacting bacteria.

In order to allow for the complexity of this microflora, some research teams screen routinely for the presence of 146 different microbial taxa in samples, while in one particularly comprehensive study, 136 distinct taxa were isolated from 38 sub-gingival samples from 22 adults. Few laboratories have the resources to monitor such a range of organisms and so the majority of studies either restrict their screening to certain, pre-selected periodontopathogens or do not discriminate many of the ill-defined or more fastidious groups of organisms. Obviously, such differences in approach create difficulties when comparing the results of particular studies. Despite these difficulties, however, certain trends have emerged. All studies agree that there is a progressive change in the composition of the microflora from health and gingivitis to periodontitis. This change involves not only the emergence of apparently previously undetected species, but also modifications to the numbers or proportions of a variety of species already present. Moreover, it proceeds in a manner which appears to be potentially highly variable both from person to person, and also at different sites within a person. Two theories have been proposed to explain the emergence of previously undetected species. It may be due to an enrichment of an organism present in only very low numbers in health due to a change in the environment during disease, or it might be due to the exogenous acquisition of periodontopathic bacteria from other diseased sites or subjects. Most studies describe complex and variable associations with bacteria without being able to define the extent to which any species or combination of species can account for the clinical difference between health, gingivitis and periodontitis. However, some of the organisms that are considered to be potentially significant aetiological agents are listed in Table 7.4.

There is a desire that a degree of specificity should exist in the microbial aetiology of periodontal diseases. As with many infections of a medical nature, it has been hoped that this specificity should be reflected in solitary or a very limited number of pathogenic species being implicated in periodontitis. Hopes that this indeed might be the case rose when it was observed that chronic periodontitis progressed not at a continuous slow rate, as was previously believed, but that there appeared to be distinct periods of disease activity over relatively short periods of time, followed by phases of quiescence or even repair. Thus, studies in which 'active lesions' had not been identified might have led to the inclusion of periodontal pockets which, at the time of sampling, were in remission and, therefore, possibly harbouring a 'non-active' microflora. This might have obscured significant associations of bacteria with those sites undergoing tissue destruction. Consequently, most recent studies have attempted to diagnose sites that are 'active' by

Table 7.4 Some bacterial species that have been commonly implicated in chronic periodontitis in adult humans

Gram-positive	Gram-negative
E. brachy	'B. forsythus'
E. nodatum	'B. pneumosintes'
E. timidium	F. nucleatum
Ps. anaerobius	P. gingivalis
Ps. micros	P. intermedia
	P. loescheii
	P. oralis

Studies have occasionally reported associations between streptococci and actinomyces with chronic periodontitis.

comparing changes in probing depth over short time periods. Probing gives a measure of the depth of the pocket, and this reflects the previous severity of disease. Despite the problems in accurately measuring probing depth, differences have been reported in the microflora from 'active' and 'inactive' periodontal sites. These differences will be discussed in a later section.

Other periodontal diseases

Acute or exaggerated forms of gingivitis can arise due to a variety of predisposing factors or circumstances. They are most commonly associated with HIV infection, pregnancy, puberty, menstruation, stress, or the use of oral contraceptives. The microbiological findings from some of these forms of gingivitis and other periodontal diseases will be considered in the following sections.

HIV-gingivitis and HIV-periodontitis

HIV-positive patients can present with an atypical gingivitis which is characterized by a band-like marginal erythema, usually accompanied by diffuse redness, which extends into the vestibular mucosa. The microflora has been determined by several groups, mainly in the USA, using conventional cultural and indirect immunofluorescence techniques. Similar trends were found: HIV-gingivitis sites were more commonly colonized by C. albicans and by a range of putative periodontopathogens including A. actinomycetemcomitans, F. nucleatum and P. gingivalis (Table 7.5), some of which are uncommon in chronic gingivitis.

Table 7.5 Percentage isolation frequency of some micro-organisms from the gingival crevice of healthy subjects, and from HIV-positive subjects with gingivitis or periodontitis

Micro-organism	HIV-negative patients (health and gingivitis)	HIV-positive patients with	
		Gingivitis	Periodontitis
A. viscosus	–*	86	70
A. actinomycetemcomitans	2	65	74
E. corrodens	4	43	30
Capnocytophaga spp.	6	29	30
F. nucleatum	31	78	66
P. gingivalis	2	86	100
P. intermedia	31	100	90
C. rectus	15	43	60
C. albicans	3	49	74

* Not reported.

HIV-positive patients can also suffer from an unusually severe and rapid form of periodontitis. The disease is generalized and is associated with blunted or cratered gingival papillae, loss of attachment, soft tissue ulceration and necrosis, as well as the more usual indicators of periodontitis (bone loss and bleeding). Indeed, there is a tendency towards spontaneous bleeding, and the condition is sometimes extremely painful. Preliminary studies suggest that the isolation frequency of a number of putative periodontopathogens such as P. gingivalis, P. intermedia, F. nucleatum and A. actinomycetemcomitans, as well as C. albicans was higher than at control sites (Table 7.5). The predominant organisms, however, were streptococci (mean proportion = 56% of the total cultivable microflora) especially S. sanguis and S. mitis, and anaerobic Gram-negative rods (24% of the total cultivable microflora). Black-pigmented anaerobes comprised over 12% of the microflora, and P. intermedia was the most prevalent species. The microflora of the HIV-gingivitis and HIV-periodontitis lesions are not markedly different, although the pathological consequences are quite distinct. This fact may be due to variations in the immune status of the respective groups.

Pregnancy gingivitis

The factors responsible for an exaggerated gingivitis in pregnancy have been investigated and linked in one study to an increase in the proportions of the black-pigmented anaerobe, *P. intermedia*, during the second trimester. Steroid hormones can be detected in GCF and following laboratory studies, the rise in numbers of *P. intermedia* was attributed to the preferential ability of this species to metabolize progesterone and oestradiol. Both of these hormones were able to replace the normal growth requirement for vitamin K by this species. However, a more recent study failed to find a significant association between *P. intermedia* and the levels of four steroid hormones in menstruating and pregnant females.

Acute streptococcal gingivitis

Acute streptococcal gingivitis is a condition affecting the gingivae which can result in severe illness. The gingivae become red, swollen and full of fluid (oedematous), the temperature is raised and the regional lymph nodes are also enlarged. If cultures are taken from the affected gingivae, a Lancefield Group A streptococcus is usually isolated. Although a number of streptococci have been implicated in the aetiology of the disease, *Streptococcus pyogenes* is usually the pathogen responsible. A number of the 'viridans-type' streptococci have also been isolated but these probably represent contamination by resident oral organisms.

This disease is usually preceded by a sore throat and hence it is possible that in the case of *S. pyogenes* there is a direct spread from throat to gingivae. One curious factor is that the disease persists for 6 to 15 days irrespective of penicillin treatment (the antibiotic of choice). It has been suggested that this persistence after therapy represents a secondary opportunistic viral infection, although as yet this is unproven.

Acute herpetic gingivitis

The majority of infectious cases of gingivitis are bacterial in origin but occasional viral gingivitis is seen, predominantly in young people. The commonest form of viral gingivitis is acute herpetic gingivitis, the causative agent of which is *Herpes simplex* type 1 (HSV–1). Many people acquire immunity to HSV–1 but do not contract acute herpetic gingivitis; thus, it is likely that this immunity is a result of primary sub-clinical infection. Acute herpetic gingivitis is seen usually in children and appears as ulcerated swellings of the gingivae which are acutely painful.

The patient's temperature is often raised and the regional lymph nodes enlarged. The symptoms may persist for 7 to 21 days and herpetic lesions may concomitantly be present on lips or any area of the oral mucosa. The diagnosis is usually made on clinical criteria although cytological smears and cytopathic effects following culture have been used for confirmation. Once infected (and this is approximately 97% of the population), then secondary herpes may occur at any time during the patient's lifetime. At present no specific treatment beyond palliative measures is available, although some new antiviral agents (e.g. acyclovir) are being evaluated. The exact pathogenesis of acute herpetic gingivitis has still to be elucidated.

Acute Necrotizing Ulcerative Gingivitis (ANUG)

Vincent's infection, or ANUG, is a painful acute condition of the gingivae. It is characterized clinically by the formation of a grey pseudomembrane on the gingivae which easily sloughs off revealing a bleeding area beneath it. ANUG can usually be diagnosed by the characteristic halitosis (bad breath) it produces. Patients suffering from ANUG are usually debilitated by another illness or are under acute emotional stress.

ANUG is a true infection and unlike chronic marginal gingivitis, micro-organisms can be seen invading the host gingival tissues. In smears of the affected tissues the invading organisms resemble spirochaetes and fusiform bacteria. Early electron microscopic investigations showed that the invading organisms consisted primarily of large and intermediate-sized spirochaetes which were present in the lesions in high numbers and in advance of other organisms. Surprisingly, there have been few bacteriological studies of ANUG using modern anaerobic techniques. Recently, in a study of 22 ulcerated sites in eight patients, a heterogeneous collection of micro-organisms were isolated. Various spirochaetes (*Treponema* spp.) were found in high numbers (approximately 40% of the total cell count) confirming the previous electron microscope studies. However, the most surprising finding, in view of the fuso-spirochaetal pattern characteristically observed, was the relatively low levels of *Fusobacterium* spp. and the high prevalence of *P. intermedia*, which averaged 3 and 24% of the total cultivable microflora, respectively (Table 7.6). Metronidazole was effective in eliminating the fuso-spirochaetal complex from infected sites and this was associated with an obvious and rapid clinical improvement. This study also concluded that disease was a result of an overgrowth of the resident microflora by obligately anaerobic species as a result of selection through the availability of host-derived nutrients in individuals who had undergone stress.

Table 7.6 Bacteriology of acute necrotizing ulcerative gingivitis

Implicated bacterial species	Mean % viable count
P. intermedia	24.0
Veillonella spp.	3.5
Fusobacterium spp.	2.6
A. odontolyticus	2.3
S. sanguis	2.4
A. viscosus	1.5
P. gingivalis	<1.0
Capnocytophaga	<1.0
	Mean % microscopic count
Treponema spp.	30.2
Large treponeme	9.9
Selenomonads	6.9
Motile rods	2.2

22 ulcerated sites in eight patients.

Juvenile periodontitis

Juvenile periodontitis is a rare condition (affecting only around 0.1% of the susceptible age group) which usually occurs in adolescents. It is thought to be a familial disease in which there is a distinct pattern of alveolar bone loss which is characteristically localized, for as yet unknown reasons, to the first permanent molars and the incisor teeth, and hence is referred to as localized juvenile periodontitis (LJP). Currently, some authorities believe that LJP may be partly due to polymorphonuclear leucocyte (polymorph) dysfunction or destruction. In a limited number of patients, failure of polymorphs to phagocytose normally has been demonstrated together with a lack of positive chemotaxis to the affected area. A contrary and interesting hypothesis is that LJP is an infectious disease due to familial transmission of *A. actinomycetemcomitans*. When this micro-organism is eliminated from a family then LJP can be arrested in some cases. These hypotheses still remain to be proven. In contrast to most other forms of periodontal disease, perhaps remarkably so in view of the aggressive nature of the tissue destruction, the plaque associated with juvenile periodontitis is sparse. Few cells are present (approximately 10^6 colony forming units per pocket) belonging to only a limited number of species. In early studies, large numbers of Gram-negative rods were isolated that were not obligately anaerobic but required carbon dioxide for growth (capnophilic). Subsequent studies identified these bacteria as *A. actinomycetemcomitans* and *Capnocytophaga* spp. This finding has important implications in treatment design because tetracycline has been shown in clinical studies to be effective in both eliminating *A. actinomycetemcomitans* from infected pockets and resolving the clinical condition. This is in contrast to other forms of chronic inflammatory periodontal disease when metronidazole would be chosen because of its specific action against obligately anaerobic bacteria.

More recent work has generally confirmed the importance of *A. actinomycetemcomitans* but failed to support the role of *Capnocytophaga* in disease. In one study of 403 subjects, 17% of healthy adults harboured low levels of *A. actinomycetemcomitans*, while 97% of localized juvenile periodontitis patients had elevated numbers of this species. Three serotypes have been described; serotype *b* strains are found more commonly in LJP patients compared with serotype *a* or *c* strains. Other evidence supporting the key role of this organism in LJP has come from treatment studies. A positive correlation was found between elimination of *A. actinomycetemcomitans* from pockets and resolution of LJP, while recurrence of the disease was related directly to the reappearance of this organism. Histological studies have shown *A. actinomycetemcomitans* invading gingival connective tissues while immunological findings have also provided additional evidence for the key role of this organism. LJP patients have high levels of antibodies in serum, saliva and GCF directed specifically against *A. actinomycetemcomitans*.

Unlike those found at healthy sites, the majority of *A. actinomycetemcomitans* strains recovered from LJP patients produce a leucotoxin, and this appears to be a major virulence factor. This leucotoxin could impair local host defences by destroying the polymorphonuclear leucocytes, especially as a major predisposing factor for juvenile periodontitis is impaired neutrophil function. Thus, in contrast to most other forms of periodontal disease, LJP appears to result from the activity of a relatively specific microflora dominated by a single species. However, as with the association of mutans streptococci with dental caries, some studies have found LJP sites in which *A. actinomycetemcomitans* is not necessarily the predominant organism; at these pockets, small spirochaetes, *E. corrodens*, *Wolinella* spp. and *F. nucleatum* are often numerous. Even when the presence of *A. actinomycetemcomitans* does correlate with LJP, its proportions ranged from 10–99% of the total pocket microflora in only 17% of 137 affected sites; in addition, 20% of sites contained from 1–10%

A. actinomycetemcomitans, while 64% of sites harboured less than 1% *A. actinomycetemcomitans*. Some sites with active breakdown have no recoverable *A. actinomycetemcomitans* implying that, in a minority of cases, the same pathological condition can be caused by other organisms.

Although juvenile periodontitis is usually localized to the specific sites described above, a severe 'generalized' form has also been described. In one of the few microbiological studies of this generalized condition, two unclassified *Treponema* species were closely associated with disease, as were *F. nucleatum*, lactobacilli, several species of *Eubacterium*, *Peptostreptococcus* spp., *P. intermedia* and *Selenomonas* spp. The significance of most of these bacteria in disease has yet to be determined.

Periodontitis in children

Destructive periodontal disease in the primary dentition (pre-pubertal periodontitis) can occur in a localized and generalized form. Localized pre-pubertal periodontitis usually has its onset before the age of five years. The gingival tissues have only minor clinical inflammation, dental plaque may be minimal, and the disease can usually be arrested by antibiotic treatment and mechanical periodontal therapy.

Generalized pre-pubertal periodontitis occurs at the time of tooth eruption and severe gingival inflammation can be common. The disease can be refractory to most treatments, and may be associated with recurrent infections and abnormalities of peripheral neutrophils and monocytes. Pre-pubertal periodontitis is rare, although in one recent study of over 2000 children in the USA, around 1% had radiographic evidence of bone loss. Two microbiological studies have found a higher prevalence of putative periodontopathogens such as *A. actinomycetemcomitans*, *P. intermedia*, *Capnocytophaga* spp. and *E. corrodens* compared to plaque taken from non-diseased children living in the same area.

Rapid-progression periodontitis

This is a poorly defined clinical condition, and it is not always clear whether such individuals could also be diagnosed as having generalized juvenile periodontitis. The amount of plaque is generally considered to be low at affected sites, which is surprising considering the rate of loss of attachment. This finding may again point to major abnormalities in the functioning of the host defences in these individuals.

Bacteria and disease progression

Bursts of disease activity have been looked for in both chronic and acute forms of periodontal disease. Two approaches have been adopted; one has been to determine the predominant organisms while the other has been to compare the presence of pre-selected putative periodontopathogens at active and inactive sites. In a longitudinal full microbiological study of eight patients with localized juvenile periodontitis, periods of apparent disease activity were reflected in a less diverse microflora in which the absolute counts and proportions of only *A. actinomycetemcomitans* and *E. corrodens* were significantly higher. Several other putative pathogens including *P. intermedia*, *F. nucleatum* and *Capnocytophaga* spp. decreased slightly in numbers or were unaffected in active pockets. In subjects described as having 'destructive periodontal disease', the isolation frequency of *P. intermedia*, '*B. forsythus*', *A. actinomycetemcomitans* and *C. rectus* (formerly *W. recta*) were significantly higher at active sites. It was equally noteworthy, however, that *F. nucleatum*, *C. gingivalis* and *E. corrodens* were recovered more commonly from active sites from some subjects but inactive sites of others. This raises the possibility as to whether (a) these species play no role in disease; or (b) a species consists of virulent and avirulent strains. Other studies of similar types of periodontal disease have come up with relatively similar groupings of bacteria (Table 7.7), although the proportions of some of the species at active sites can be low (<5% of the total cultivable microflora), while their prevalence and range of viable counts at active and inactive sites can overlap considerably. Indeed, in the largest comparison of this type (170 sites in 76 subjects), a heterogeneous collection of organisms was recovered, in which only *F. nucleatum* showed a reasonable association with disease activity, which suggested that many microbiologically-distinct forms of chronic periodontal disease might exist.

In the second approach, the prevalence of three putative periodontopathogens at sites with active or inactive disease was determined. One or more of the species *A. actinomycetemcomitans*, *P. gingivalis* or *P. intermedia* were isolated from 99% of progressing pockets but from only 40% of inactive sites. It was hoped that this might lead to a predictive test to identify sites at risk of future disease progression. When all three species were absent from a site, the probability of the pocket undergoing future progression was only 1%, but in a prospective study, only 20% of sites suffered attachment loss when one or more of these organisms were present. Similar studies, in which only *P. gingivalis* was monitored, found that this species was uncommon at non-diseased sites, and periodontal breakdown was more likely when it was present. Experimental evidence that *P. gingivalis* has the potential to cause bursts of tissue

Table 7.7 Bacteria that have been associated with disease activity in periodontal lesions

Periodontal disease	Bacteria
Localized juvenile periodontitis	*A. actinomycetemcomitans* *E. corrodens*
Localized and generalized destructive periodontal disease	*F. nucleatum* (miscellaneous species)
Destructive periodontal disease	*1 *P. gingivalis* 'B. forsythus' *C. rectus* Small spirochaetes
	2 *A. actinomycetemcomitans* *P. intermedia* 'B. forsythus' *C. rectus* *S. sputigena*
	3 *C. rectus* 'B. gracilis' *E. corrodens*
Advanced peridontal disease	*P. intermedia* *P. gingivalis* *A. actinomycetemcomitans*
Recurrent periodontitis	'B. forsythus'

* Different studies.

destruction has come from studies using a monkey periodontitis model. In spite of the presence of an existing diverse microflora, significant bone loss occurred almost exclusively at sites implanted with *P. gingivalis*. Thus, as with the situation with mutans streptococci and dental caries, there is more chance of ruling out the possibility of disease when key organisms are absent than predicting future episodes of disease when they are present.

Summary

Periodontal diseases are generally associated with diverse mixed cultures of predominantly obligately anaerobic bacteria, many of which are Gram-negative and asaccharolytic. An exception to this is some of the more acute or rapid forms of periodontal disease, such as LJP, where capnophilic species such as *A. actinomycetemcomitans* show a close association with attachment loss. Acute forms of periodontal disease may also reflect a mis-functioning of the host defences. In chronic periodontitis, the composition of the microflora can vary with the depth of the pocket and with the severity of the disease. It also appears that microfloras with a radically-dissimilar composition can produce an apparently similar pathological condition. However, this variation may merely reflect the lack of precision in clinical diagnosis or sample taking, or both. It may also reflect differences in the level of disease activity in the pocket at the time of sampling.

THE HABITAT AND SOURCE OF PERIODONTOPATHIC BACTERIA

The predominant bacteria found in the various types of periodontal disease are different to those that are prevalent in the healthy gingival crevice. One of the most intriguing questions in periodontology, therefore, concerns the reservoir or natural habitat of periodontopathogens. Usually these bacteria are found only occasionally and in low numbers in the healthy gingival crevice. It is possible that they might be more widespread, but in numbers below those capable of detection by conventional techniques. However, if there is a change in the local environment, for example, as a result of trauma, an alteration in the immune status of the host, or an increase in gingival crevicular fluid flow following plaque accumulation due to poor oral hygiene, then the growth of periodontopathic bacteria might be favoured at the expense of other species. This could lead to a shift in the proportions of the resident sub-gingival microflora in an analogous way to the increases in mutans streptococci and *Lactobacillus* spp. seen prior to caries development following the repeated ingestion of dietary carbohydrates. Evidence for this has come from laboratory studies in which sub-gingival plaque was passaged repeatedly through human serum (as a substitute for GCF). Eventually the microflora became dominated by species associated with periodontal destruction, such as black-pigmented anaerobes, peptostreptococci, *Fusobacterium* spp. and spirochaetes, many of which were not detected in high numbers in the original plaque samples. Likewise, *in vitro* studies have shown that increasing the pH from 7.0 to 7.5 (as can occur during inflammation) can allow *P. gingivalis* to rise from <1% to >99% of a microbial community of black-pigmented anaerobes. If this situation does occur in a pocket, periodontal diseases could be regarded as endogenous infections caused by an imbalance in the composition of the resident microflora at a site due to an alteration in the ecology of the crevice.

Some periodontopathic bacteria can also attach to mucosal surfaces, and recent studies of human volunteers have isolated *P. melaninogenica*, *P. intermedia* and *Capnocytophaga* from the dorsum of the tongue and from tonsils. Spirochaetes, *P. denticola* and various motile organisms were also recovered from tonsils while *Fusobacterium* spp. were found on the tongue, tonsils and buccal mucosa. In contrast, *A. actinomycetem-comitans* was never isolated from any mucosal surface. Following an experimental gingivitis study with these volunteers, in which the plaque microflora from diseased sites was compared with the mucosal carriage of the above bacteria, it was proposed that the dorsum of the tongue may act as a nidus for certain periodontopathic bacteria.

It has also been suggested that some forms of periodontal disease should be regarded as exogenous infections because the causative bacteria (e.g. *P. gingivalis*) are not widely distributed in the normal mouth and should not, therefore, be regarded as members of the resident microflora. Intriguingly, there has been a single recent report of a child with Papillon-Lefèvre syndrome who was consistently re-infected with *A. actinomycetemcomitans*. When *Actinobacillus* strains from possible sources of infection were compared using restriction endonuclease mapping of the bacterial DNA, it appeared that the child had acquired its biotype of *A. actinomycetemcomitans* from the family pet dog. Further studies will be necessary to determine whether other episodes of *Actinobacillus*-associated periodontal destruction in children also represent a zoonosis. Treatment might, therefore, also involve elimination or suppression of putative periodontopathogens from their primary oral reservoirs.

TREATMENT AND PREVENTION OF PERIODONTAL DISEASES

Plaque control

Plaque control is fundamental to (a) the prevention of gingivitis; (b) treatment of established disease (gingivitis or periodontitis); and (c) in the maintenance of health following effective treatment. Plaque control can be achieved by conventional oral hygiene measures such as tooth brushing and flossing, which can be augmented by professional prophylaxis during routine visits to the dentist. As discussed in Chapter 6, a number of antimicrobial agents have been incorporated into toothpastes and mouthwashes which have anti-plaque and anti-gingivitis benefits. These agents include metal salts (e.g. zinc citrate), enzymes (e.g. glucose oxidase/amyloglucosidase), plant extracts (e.g. sanguinarine), bisbiguanides (e.g. chlorhexidine) and phenols (e.g. Triclosan). These antimicrobial agents can not only inhibit the growth of relevant sub-gingival bacteria, but can interfere with the expression of virulence determinants (e.g. protease activity) when present at sub-MIC levels. Their efficacy in preventing gingivitis has been demonstrated using the experimental gingivitis model in human volunteers. Mouthwashes containing, for example, chlorhexidine are also effective against established gingivitis; it is yet to be determined how effective antimicrobial agents delivered from toothpastes are in resolving pre-existing disease.

In more advanced forms of periodontal disease, agents delivered from mouthwashes and toothpaste are usually unable to penetrate sufficiently into the pocket to be effective. In this situation, treatment requires professional plaque control which in some circumstances may require surgery so that access to the root surface is achieved. The depth of the pocket will determine the complexity of the surgical treatment. In extreme cases, not only is there a need to remove plaque and/or calculus, which may be firmly attached to the root, but also the outer surface layers or cementum (root planing). The latter is advocated by some because of the possible penetration into cementum of cytotoxic or inflammatory products of sub-gingival bacteria, especially lipopolysaccharide (LPS). Even after thorough root planing to remove obvious deposits of plaque and calculus, residual bacteria may still be present, and sites can be repopulated rapidly leading to further loss of attachment in some pockets. Consequently, post-surgical control of microorganisms is sometimes necessary. This again can involve meticulous supragingival plaque control (to reduce the likelihood of sub-gingival colonization) or the use of antimicrobial agents, such as chlorhexidine, or even systemic antibiotics, such as tetracycline, amoxycillin, or metronidazole. Antibiotics should only be used in special circumstances, such as localized juvenile periodontitis or in refractory periodontal disease, because of the problems of antibiotic resistance. A preferred approach is to apply antimicrobial agents locally, for example, using hollow fibres impregnated with the drug of choice, by direct irrigation of the pocket, or by inserting slow release materials, such as acrylic or ethyl cellulose polymers. Chlorhexidine, metronidazole and tetracycline have been successfully delivered in this way.

Refractory peridontal disease

In a small minority of cases, pockets with destructive periodontal disease fail to respond to conventional treatment even when augmented with antibiotic therapy. In one study of 13 subjects with refractory sites, the microbiological findings indicated that these pockets did not form a homogeneous group in terms of the predominant species that were recovered. Three major groupings of bacteria were found (Table 7.8); *E. corrodens* was also found at some refractory sites. Elevated serum antibody levels against *A. actinomycetemcomitans*, *F. nucleatum*, *P. gingivalis*, *P. oralis*, *P. intermedia*, *E. corrodens* and *C. concisus* were also observed.

Table 7.8 Predominant bacteria found in refractory peridontal disease in 13 subjects

Group*	Number of subjects	Bacteria
1	3	'B. forsythus', F. nucleatum, C. rectus
2	3	S. intermedius, P. gingivalis, Ps. micros
3	7	S. intermedius, F. nucleatum ± P. gingivalis

* Three major groupings of bacteria were detected.
  E. corrodens was also found at some refractory sites.

One reason for a pocket to be refractory might be if mechanical plaque control was inadequate during periodontal therapy. Another reason might be due to the development of antibiotic resistance among the members of the pocket microflora. In a large cross-sectional study of 500 patients with refractory sites, nearly one third of pockets were colonized by non-resident oral micro-organisms including yeasts, enterobacteria and *Pseudomonas aeruginosa*. The majority of yeasts were *C. albicans* while the enterobacteria included *Proteus mirabilis*, *Klebsiella* spp. and *Enterobacter* spp.; when present these organisms could constitute between 20–40% of the microflora. All yeasts and the majority of enteric rods and pseudomonads were resistant to tetracycline, penicillin G, and erythromycin. It may be necessary, therefore, to perform a microbiological screening on refractory patients prior to treatment with antibiotics.

PATHOGENIC DETERMINANTS OF PERIODONTOPATHIC BACTERIA

Bacterial invasion of the gingival tissues

Despite the frequent massive presence of bacteria in the periodontal pocket, microbial invasion of the host tissues appears to be rare. An exception to this is in acute necrotizing ulcerative gingivitis where there is a consistent (but superficial) invasion of the gingival connective tissues by spirochaetes. It has also been reported recently that bacteria can invade tissues in other acute forms of periodontal disease, e.g. localized juvenile periodontitis, in the late stages of severe chronic periodontitis, and in HIV-associated periodontal disease. The bacteria have been detected using scanning and transmission electron microscopy of sectioned tissue while immunocytological techniques have been used in some studies to identify the invading organisms. In juvenile periodontitis, invasion of both gingival epithelium and adjacent connective tissues has been described; the organisms that have been described include numerous coccobacillary-shaped bacteria and *Mycoplasma*-like organisms. Other studies claim to have specifically identified cells of *A. actinomycetemcomitans* in the host tissues.

In advanced cases of periodontitis, tissue invasion by a number of bacterial morphotypes has been seen in some but not all of the sites examined. Most bacteria appear to be Gram-negative and include cocci, rods, filaments and spiral-shaped cells, although some Gram-positive bacteria have also been observed. The micro-organisms are usually located in enlarged epithelial intercellular spaces but when found in the underlying connective tissue there is evidence of severe (host) cell damage and collagen breakdown. It should be stressed, however, that, although bacterial *invasion* has been observed, it is not a common occurrence and may sometimes only represent bacterial *translocation* (passive entry) into the gingival tissues instead. Moreover, it can be extremely difficult to locate bacteria in the tissues and considerable care has to be taken not to 'contaminate' any tissue specimen with pocket bacteria during its sampling and processing.

Potential mechanisms of tissue damage

As the major forms of periodontal disease are characterized by the progressive destruction of the supporting tissues of teeth in the apparent absence of significant tissue invasion, tissue damage must be mediated primarily by surface components and extracellular products of bacteria (Table 7.9). These bacterial products might cause destruction of gingival tissue by two mechanisms. In one, damage is believed to result from the direct action of bacterial enzymes and cytotoxic products of metabolism on the tissues. In the other, bacterial components are only indirectly responsible, causing tissue destruction as the inevitable side effect of the protective host inflammatory response to the plaque antigens.

Table 7.9 Bacterial factors implicated in the aetiology of peridontal diseases

Stage of disease	Bacterial factor
Attachment to host tissues	Surface components, e.g. 'adhesins' Surface structures, e.g. fimbriae, fibrils
Multiplication at a susceptible site	Protease production to obtain nutrients Development of food chains Inhibitor production, e.g. bacteriocins
Evasion of host defences	Capsules and slimes PMN-receptor blockers Leucotoxin Immunoglobin-specific proteases Complement-degrading proteases Suppressor T cell induction

Tissue damage (a) direct	**Enzymes**
	  'Trypsin-like' protease
	  Collagenase
	  Hyaluronidase
	  Chondroitin sulphatase
	Bone resorbing factors
	  Lipoteichoic acid
	  Lipopolysaccharide
	  Capsule
	Cytotoxins
	  Butyric and propionic acids
	  Indole
	  Amines
	  Ammonia
	  Volatile sulphur compounds

(b) indirect	Inflammatory response to plaque antigens Interleukin–1 production and proteinase synthesis in response to plaque antigens

Direct pathogenicity

A range of species, but particularly *P. gingivalis*, '*B. forsythus*', *P. intermedia*, *Capnocytophaga* spp., *A. actinomycetemcomitans*, *Peptostreptococcus* spp. and *Treponema* spp. synthesize between them a number of cell surface or extracellular enzymes that could potentially weaken the integrity of the periodontal tissues. The enzymes include collagenase, glycylprolyl peptidase, hyaluronidase, a trypsin-like protease, and chondroitin sulphatase. In Gram-negative species, these enzymes are also located on outer membrane vesicles that are shed from the bacterial cell surface during growth, enhancing the likelihood of tissue penetration by these enzymes. Once the integrity of the tissues is impaired, further damage might arise from the increased penetration of cytotoxic bacterial metabolites such as indole, amines, ammonia, volatile sulphur compounds (e.g. methyl mercaptans, H_2S), and butyric and propionic acids. *Porphyromonas gingivalis* has been shown to have the greatest proteolytic activity of the Gram-negative bacteria isolated in high numbers from sites affected by periodontal diseases, and was also the most virulent species when inoculated into animals in a simple pathogenicity test. Other components able to cause direct tissue damage include surface antigens, such as the lipoteichoic acid of some Gram-positive bacteria and endotoxin (lipopolysaccharide, LPS) of Gram-negative cells. LPS has been shown in laboratory studies to stimulate bone resorption. Molecules of even greater biological activity are being identified in some Gram-negative bacteria. One example of this is the capsule of *A. actinomycetemcomitans* which is 1000-fold more potent at causing bone resorption *in vitro* than the purified LPS of this organism.

The production of some of these pathogenic determinants can be markedly influenced by environmental conditions that are relevant to changes that occur during the transition from a normal gingival crevice to a periodontal pocket. Using a simple mouse pathogenicity test, cells of *P. gingivalis* grown with an excess of haemin (an essential cofactor for growth that would be obtained *in vivo* from the degradation of host haeme-containing molecules present in GCF) were more virulent than cells grown haemin-limited. This was associated with 3-fold higher levels of the trypsin-like protease in the virulent cells. A rise in the pH to values known to occur during inflammation was found to enhance the growth of *P. gingivalis* (optimum pH 7.5–8.0; Chapter 2) and increase the activity of the trypsin-like protease. Increasing the growth rate of *P. gingivalis* has also been found to increase the activity of the trypsin-like protease, while slow growth rates and haemin-restricted conditions appear to induce vesicle production. This may be a response by the organism to conditions of stress; proteases located on the surface of the vesicle may liberate fresh nutrients from nearby sources.

A great deal of attention has been focused on the trypsin-like protease of *P. gingivalis*. This enzyme has proved difficult to characterize biochemically but, except for aspects of its substrate specificity, it is dissimilar to classical mammalian trypsin enzymes. Consequently, it has been renamed by one group as 'gingivain', because it has properties more consistent with those of a cysteine-proteinase. This enzyme has many properties of relevance: it can degrade a number of host proteins (albumin, transferrin, haptoglobin, haemopexin, complement, immunoglobulins) some of which are involved in the operation of the host defences. It can also contribute to the destruction of tissue components including fibronectin and collagen, and it can cause the direct lysis of red blood cells, a process that may enable *P. gingivalis* to obtain protohaem *in vivo*. Whether this enzyme also acts as the haemagglutinin, whereby cells of *P. gingivalis* can haemagglutinate red blood cells, remains to be determined.

No single property has been found to confer pathogenicity to any putative periodontopathogen. Studies of an avirulent mutant of *P. gingivalis* have shown it to possess three-fold lower trypsin-like protease activities but to have a greater potential to degrade collagen; the virulent parent strain also possessed a thicker capsule which may enable these cells to avoid phagocytosis. The virulence of *A. actinomycetemcomitans* is strongly associated both with the production of a leucotoxin and the presence of a capsule.

Indirect pathogenicity

In indirect pathogenicity, any sub-gingival plaque bacterium could be considered to be playing a role in tissue destruction if they contribute to an inflammatory host response. Such a response could lead to tissue destruction being mediated by, for example, proteases released from neutrophils or by enzymes released from connective tissue cells. The latter may arise due to the production of cytokines, such as interleukin–1, following the interaction of a range of non-specific bacterial antigens with inflammatory cells. Interleukin–1 can induce the release of host enzymes, such as collagenase and stromelysin, that degrade the connective tissue matrix. It is likely that both the direct and the indirect mechanisms of tissue damage operate *in vivo*.

PATHOGENIC SYNERGISM AND PERIODONTAL DISEASE

One of the most consistent and controversial features of the micro-biology of periodontal diseases is the isolation of complex mixtures of bacteria from diseased sites. In particular, in chronic periodontitis, the composition of these mixtures can differ considerably both between and within studies of patients presenting with apparently similar clinical features. These variations might be explained by (a) differences in sampling and plaque-processing methods; (b) difficulties in accurately diagnosing the clinical condition; or (c) plaque being sampled during both 'active' and 'inactive' phases of the disease. However, for the establishment of disease, an organism must gain access to and adhere at a susceptible site, multiply, overcome or evade the host defences, and then produce or induce tissue damage. A large number of virulence traits are needed, therefore, for each stage in the disease process (Table 7.9), and it is unlikely that any single organism will produce all of these factors optimally or in every situation.

An alternative explanation for some of the observed variations in microflora associated with periodontal disease could be that tissue destruction is a result of consortia of interacting bacteria. In this way, periodontal diseases are a particularly striking example of a synergistic infection whereby organisms that are individually unable to satisfy all of the requirements necessary to cause disease combine forces to do so. Thus, although only a few species (e.g. *P. gingivalis*, *P. intermedia*, *Treponema* spp.) produce enzymes that cause tissue damage directly, the persistence of these 'primary pathogens' in the pocket may be dependent on a number of organisms to provide means of attachment (e.g. receptors for coaggregation on *Streptococcus* and *Actinomyces* spp.) or essential nutrients for growth (e.g. vitamin K, protohaeme, succinate; Figure 5.9). Similarly, the bacteria that support the growth of the 'primary pathogens' may also require other organisms to suppress or inactivate the host defences, or to inhibit competing organisms (e.g. by bacteriocin production) to ensure their survival. Bacteria could also have more than one function in the aetiology of periodontal disease and a schematic diagram illustrating this pathogenic synergism is shown in Figure 7.3. Some evidence to support this proposal is provided by infection studies in animals and recent laboratory work. Transmissible infections were consistently produced only when a black-pigmented anaerobe was present in a mixed culture inoculum. The other organisms (*P. oralis*, *V. parvula*, anaerobic streptococci, facultatively anaerobic Gram-positive rods) were non-infective and their function appeared to be merely to provide essential growth factors for the black-pigmented anaerobe. Similarly, in studies of bacteria isolated from periodontal

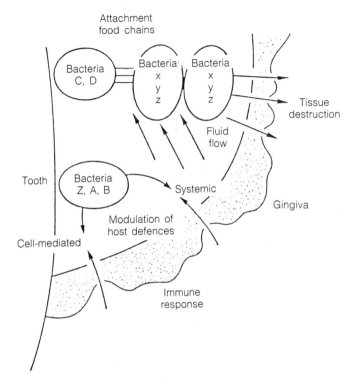

Figure 7.3 Pathogenic synergy in the aetiology of periodontal diseases. Bacteria capable of causing tissue damage directly (e.g. species X, Y and Z) may be dependent on the presence of other cells (e.g. organisms C and D) for essential nutrients or attachment sites, so that they can grow and resist the removal forces provided by the flow of gingival crevicular fluid. Similarly, both of these groups of bacteria may be reliant for their survival on other organisms (Z, A, B) to modulate the host defences. Individual bacteria may have more than one role (e.g. organism Z) in the aetiology of disease.

pockets, it was found that the production of protohaeme by *C. rectus* (formerly *W. recta*) stimulated the growth of *P. gingivalis*, while the growth of *C. rectus* was itself enhanced by formate produced by *P. melaninogenica*. Thus, our ability to interpret results from future microbiological studies of periodontal disease would be greatly enhanced if we knew more about the role (or niche; Chapter 1) of particular species in the disease process. An organism could still be highly significant in disease without having the potential to cause tissue destruction directly, while in other pockets, different bacteria could fill identical roles.

SUMMARY

Periodontal diseases are a group of disorders that affect the supporting tissues of the teeth. In later life, more teeth are lost due to periodontal diseases than dental caries. The predominant microflora found in disease differs from that in health, but there is no single or unique pathogen. Most of the bacteria associated with disease are Gram-negative and obligately anaerobic, except for localized juvenile periodontitis, where the microflora is mainly capnophilic. Although the microflora in disease is diverse, certain species are commonly found at sites undergoing tissue breakdown; these include *P. gingivalis*, *P. intermedia*, *A. actinomycetemcomitans*, '*B. forsythus*', *C. rectus* (formerly *W. recta*), *F. nucleatum*, and spirochaetes. Many of these species are highly proteolytic and can degrade host tissues and/or components of the host defences. Bacterial invasion of tissues is rare except in some acute conditions such as ANUG and localized juvenile periodontitis. Acute forms of periodontal disease may also be due to abnormalities in the functioning of the host defences. Tissue destruction is generally mediated by cell surface proteases and extracellular cytotoxic compounds. Organisms can evade the host defences by the action of specific proteases or the presence of a capsule. Periodontal diseases involve the destruction of tissues directly by bacterial enzymes and indirectly as a consequence of the host inflammatory response. Treatment and prevention involve good oral hygiene which may be augmented by the use of antimicrobial agents.

FURTHER READING

Addy, M. (1990) Chemical plaque control. In *Periodontics. A Practical Approach*. (Ed. B. Kieser) Wright, London, pp. 527–34.

Frisken, K. W., Tagg, J. R., Laws, A. J. and Orr, M. B. (1987) Suspected periodontopathic micro-organisms and their oral habitats in young children. *Oral Microbiology and Immunology*, **2**, 60–4.

Johnson, N. W. (ed.) (1991) *Risk Markers for Oral Diseases. Vol. 3. Periodontal Diseases: Markers and disease susceptibility and activity*. Cambridge University Press, Cambridge.

Loesche, W. J., Syed, S. A., Laughon, B. E. and Stoll, J. (1982) The bacteriology of acute necrotizing ulcerative gingivitis. *J. Periodontology*, **53**, 223–30.

Maiden, M. F. J., Carman, R. J., Curtis, M. A., Gillett, I. R., Griffiths, G. S., Sterne, J. A. C., Wilton, J. M. A. and Johnson, N. W. (1990) Detection of high risk groups and individuals for periodontal diseases: laboratory markers based on the microbiological analysis of subgingival plaque. *Journal of Clinical Periodontology*, **17**, 1–13.

Mayrand, D. and Holt, S. C. (1988) Biology of asaccharolytic black-pigmented *Bacteroides* species. *Microbiological Reviews*, **52**, 134–52.

Meikle, M. C., Heath, J. K. and Reynolds, J. J. (1986) Advances in understanding cell interactions in tissue resorption. Relevance to the pathogenesis of periodontal diseases and a new hypothesis. *Journal of Oral Pathology*, **15**, 239–50.

Moore, L. V. H., Moore, W. E. C., Cato, E. P., Smibert, R. M., Burmeister, J. A., Best, A. M. and Ranney, R. R. (1987) Bacteriology of human gingivitis. *Journal of Dental Research*, **66**, 989–95.

Moore, W. E. C., Holdeman, L. V., Cato, E. P., Smibert, R. M., Burmeister, J. A., Palcanis, K. G. and Ranney, R. R. (1985) Comparative bacteriology of juvenile periodontitis. *Infection and Immunity*, **48**, 507–19.

Moore, W. E. C., Holdeman, L. V., Cato, E. P., Smibert, R. M., Burmeister, J. A. and Ranney, R. R. (1983) Bacteriology of moderate (chronic) periodontitis in mature adult humans. *Infection and Immunity*, **42**, 510–15.

Socransky, S. S., Haffajee, A. D., Smith, G. L. F. and Dzink, J. L. (1987) Difficulties encountered in the search for the etiologic agents of destructive periodontal diseases. *Journal of Clinical Periodontology*, **14**, 588–93.

Zambon, J. J. (1985) *Actinobacillus actinomycetemcomitans* in human periodontal disease. *Journal of Clinical Periodontology*, **12**, 1–20.

8 The role of oral bacteria in other infections

In Chapters 6 and 7, the role of micro-organisms in the two main oral diseases, caries and periodontal disease, was discussed. Micro-organisms resident in the oral cavity can cause a variety of other diseases and this chapter will discuss the commonest of these. In previous chapters, the importance of homeostasis in maintaining the stability of oral microflora was emphasized. If the homeostasis of the oral microflora is lost then **opportunistic infections** can occur. Opportunistic infections are derived from micro-organisms present in the host's own microflora, i.e. they are *endogenous* infections. Microbial imbalance and loss of homeostasis has been described in plaque (Chapters 6 and 7), but oral mucosal surfaces can also suffer from endogenous infections following the breakdown of microbial homeostasis.

In contrast, infections that are caused by micro-organisms not normally found in the oral microflora are termed **exogenous. Frank pathogens** (e.g. *Mycobacterium tuberculosis*) are those micro-organisms which when present always cause infection; this type of infection is rare in the oral cavity.

PREDISPOSING FACTORS

Major perturbations to the composition of the oral microflora often result in infections. These major perturbations cause selective overgrowth of one or more components of the microflora. A definite number of a particular micro-organism is needed to cause infection; this number is called the **minimum infective dose**; once exceeded, the infection is likely to occur. An alternative way in which infections occur is by the introduction of exogenous micro-organisms into a microflora which has lost homeostasis. The introduction and growth of exogenous micro-organisms is usually prevented by the resident microflora having colonization resistance (Chapter 2). A large number of predisposing factors can result in oral infections; some of these are listed in Table 8.1. Also included in Table 8.1 are the effects of these predisposing factors on hosts' defence mechanisms. Often the effects of these predisposing factors can be temporary (e.g. following short-term antibiotic therapy), but where the factor is not reversible (e.g. loss of salivary secretion) then the effect on the microflora is permanent. In such circumstances the host may be rendered permanently susceptible to opportunistic infection. In the treatment of all opportunistic oral infections the predisposing factor must be eliminated or corrected if a permanent cure is to result.

Table 8.1 Predisposing factors which result in oral infections

Predisposing factors	Possible effect on defence mechanism	Oral infection
Physiological		
Old age and infancy	Diminution of immunoglobulin, decrease in salivary flow	Candidosis Root caries
Pregnancy	Unknown	Gingivitis
Trauma		
Local	Loss of tissue integrity	Various opportunistic infections
General	General debilitation, dehydration	Candidosis
Malnutrition	Deficiencies of iron, folate	Candidosis
Endocrine disorders	Mostly unknown	Fungal infections
AIDS	Reduction of host immune defences	*Candida*. Various opportunistic oral infections
Antibiotic therapy	Loss of colonization resistance, selection of resistant microflora	Candidosis, various opportunistic infections
Chemotherapy	Xerostomia, local mucosal effect	Candidosis, caries
Oral malignancies	Xerostomia, loss of muscular function	Caries, Candidosis

ABSCESSES IN THE ORAL REGION

When the predisposing factors allow, micro-organisms can multiply and form abscesses. An abscess is a localized collection of pus and may be either acute or chronic. Pus is composed of micro-organisms and their products, inflammatory cells, tissue breakdown products, proteins from serum and other organic material. Abscesses cause local tissue destruction, usually through damage exerted by pressure or by the action of the enzymes they contain. These enzymes can be derived either directly from the bacteria or from the inflammatory response of the host. The treatment of abscesses is primarily by surgical drainage of the pus. Surgery destroys the integrity of the abscess and allows the host's defences to resolve the infection. Antimicrobial agents are used as an adjunct to the surgical treatment.

Abscesses occur where micro-organisms are able to multiply in a confined space. This may occur, for example, at the apex of a tooth root where a periapical abscess may result. Acute abscesses may drain by destroying soft tissue until the pus is released from the tissue and drains either into the mouth or extraorally. If the cause of the abscess is not removed, the place where the pus drains may get partially lined by oral mucosa and form a sinus. Most oral abscesses result in some bacteria entering the bloodstream (bacteraemia). Usually this is not significant to the host as phagocytic cells, such as polymorphonuclear leucocytes or macrophages, together with other haematological antimicrobial factors, destroy the bacteria. If the bacterial release into the blood stream is large then they may divide (septicaemia); this can cause loss of consciousness and shock. Another consequence of large numbers of bacteria in the blood stream is the infection of an organ essential for life. Abscesses of the brain, liver and kidney have been reported from oral infections. If an abscess does not resolve, then other damage or secondary infections may result. This is particularly important in the heart where endocarditis may occur as a result of a bacteraemia derived from the original abscess. The majority of abscesses are bacterial, but fungi (usually *Candida* spp.) can also be the cause. The possible results of an abscess in the oral cavity are shown in Figure 8.1.

DENTAL ABSCESSES

The dental abscess (synonymous terms include the periapical or dentoalveolar abscess) is a collection of pus in the pulp or around the root of a tooth. Dental abscesses usually result from necrosis of the pulp following the progression of dental caries. There are other causes of dental abscesses besides caries and these are shown in Figure 8.2.

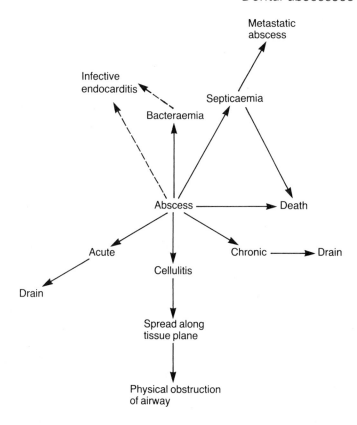

Figure 8.1 The possible result of an abscess in the oral cavity.

A dental abscess may result from either exogenous or endogenous micro-organisms. The exogenous micro-organisms involved depend on their source, but staphylococci, either *Staph. aureus* or various coagulase-negative staphylococci, are common. The micro-organisms involved in dentoalveolar abscesses have been the subject of a considerable amount of study recently. It has been found that the method of sampling and culturing the micro-organisms is of prime importance. The use of aspirated pus as opposed to swabs has been shown to increase the qualitative and quantitative recovery of bacteria from these abscesses. The use of pre-reduced media and anaerobic cabinets has enhanced the recovery and isolation of the obligate and facultative anaerobes. Dental abscesses usually contain a mixture of facultative and strictly anaerobic micro-organisms; single species of micro-organism causing an abscess are usually rare. The type and diversity of micro-organism is shown in Table 8.2. Oral streptococci are often recovered from dental abscesses

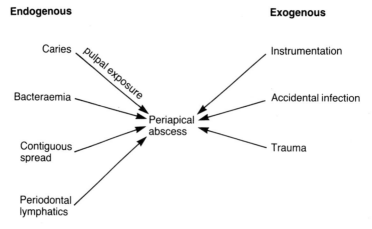

Endogenous   **Exogenous**

Figure 8.2 Scheme for the routes of infection resulting in a periapical abscess.

and are present in the resident plaque microflora. This has led some research workers to speculate that the pathogenesis of dental abscesses initially involves facultative oral streptococci. As the abscess develops there is a progression from a facultative to an anaerobic state. Such a hypothesis is logical, as it can be postulated that as the tooth pulp dies the Eh is lowered, thereby favouring the growth of obligate anaerobes. These increase in number and grow in the presence of serum or tissue-derived products which may diffuse into the abscess. In the serum are host proteins which can be hydrolyzed to peptides by proteolysis by anaerobes. The presence of mixtures of facultatively and obligately anaerobic bacteria may enhance abscess formation beyond that of the individual micro-organisms; this is another example of pathogenic synergism (Chapter 7). The essential treatment of oral abscesses is surgical drainage of the pus; without surgery, resolution is unlikely to occur. Antibiotics such as the penicillin derivatives and metronidazole are

Table 8.2 Bacteria commonly isolated from dental abscesses

Facultative anaerobes	Obligate anaerobes
S. milleri-group (S. anginosus)	Anaerobic cocci
S. oralis-group	Prevotella oralis
Mutans streptococci	Peptostreptococcus spp.
Lactobacillus	Porphyromonas gingivalis
Actinomyces spp.	Veillonella
Haemophilus spp.	Non-pigmented Prevotella spp.

useful adjuncts to surgical treatment, particularly if systemic involvement (raised temperature) is present.

ABSCESSES ASSOCIATED WITH OSSEOINTEGRATED IMPLANTS

The last decade has seen the development of oral metallic or metalloceramic implants to replace dental tissue. The implants, if carefully placed, can integrate with the underlying bone. If the surgical or aseptic technique during insertion of the implant is poor then infection results. Infections that arise immediately after implantation are usually due to staphylococci, either *Staph. aureus* or the coagulase-negative species. The bacteria responsible for infections that arise after initial implantation are usually similar to those associated with dental abscesses, and various oral streptococci and *Prevotella* spp. are involved. Treatment of the abscesses is usually by draining the pus and removing the implant; the latter may be difficult if some osseointegration has occurred. Antibiotics such as the penicillin derivatives or metronidazole are useful if systemic involvement has occurred.

OSTEOMYELITIS

Infections of the bone that are either chronic or acute are called osteomyelitis. The incidence of osteomyelitis in Western civilizations has decreased considerably in recent years, although the exact reason is not clear. It is now rare for osteomyelitis to occur following an extensive dental abscess, an occurrence not uncommon a quarter of a century ago. The other predisposing factor that can lead to osteomyelitis is severe trauma caused by fracture of the jaws with exposed bones (e.g. road traffic accidents) but even this is not common. The micro-organisms that have usually been implicated in this condition are the staphylococci, such as *Staph. aureus* and other coagulase-negative species. Recent studies have shown that the micro-organisms actually involved in osteomyelitis are Gram-negative bacteria, such as black-pigmented anaerobes (including *Prevotella* spp.), and *Fusobacterium* spp. usually as a mixture. The presence of these obligate anaerobes may be a reflection of the avascular nature of the lesion and the consequent low redox potential in the bone.

In recent years another type of osteomyelitis has been recognized. When the jaws are irradiated for neoplastic (cancerous) conditions, they may develop bone necrosis, which becomes avascular due to the therapy. In these conditions the oral microflora also changes considerably, mainly as a result of the loss of function of the salivary glands and subsequent xerostomia. Exogenous Gram-negative bacteria, such as *Escherichia coli*, pseudomonads, *Proteus*, and *Klebsiella* can colonize and infect the bone. Recent studies have also shown that this colonization is associated with the release of large quantitites of endotoxin which is probably responsible for oral ulceration and exposure of bone. The osteomyelitis seen in patients with AIDS is often of this type. Such extreme examples of osteomyelitis are not common, but they reinforce the role of saliva in the regulation of the oral microflora, particularly in the prevention of colonization by exogenous Gram-negative bacteria. The treatment of this type of osteomyelitis is usually with high doses of antibiotics, such as the penicillinase-resistant derivatives of penicillin and the cephalosporins. Surgery is often necessary to remove the spicules of bone (sequestrae) to help healing.

DRY SOCKETS

One specialized form of localized osteomyelitis that occurs in the oral cavity is the 'dry' socket. This, as its name implies, is a tooth socket which fails to heal after an extraction. The micro-organisms that can be isolated from dry sockets are sparse and are often obligately anaerobic. *Prevotella* spp. and *Fusobacterium* spp. have been isolated, but so also have *Staph. aureus*, *S. pyogenes* and *Actinomyces* spp. There is still some argument as to whether this is an infectious lesion or whether it is due to severe local trauma. Some dry sockets improve with the administration of metronidazole which would implicate obligately anaerobic species in the pathology of this condition. The incidence of dry sockets is considerably reduced by the preoperative use of chlorhexidine subgingivally before extraction; again, this is evidence of an infective aetiology. Dry sockets are acutely painful and are treated by removing debris and placing pain-relieving antiseptic dressings in the socket.

ACTINOMYCOSIS

Actinomycotic infections are usually chronic, long-standing infections of the head and neck. There is usually a history of mild trauma, e.g. tooth extraction, or a blow to the jaw. Classically, actinomycotic lesions present as a chronic abscess, usually at the angle of lower jaw, with multiple external sinuses. In some cases a thick fluid can be expressed from the sinuses which contains yellow, granular, particulate 'matter', often referred to as 'sulphur granules'. These granules are aggregates of actinomyces filaments and may inhibit penetration of antibiotics. Actinomycosis rarely penetrates bone and usually affects the lower jaw and not the upper jaw. The slow growth of the *Actinomyces* spp. induces a granulomatous reaction around the lesions which effectively 'walls' it off. This is important in therapy as these fibrous walls must be broken down if the lesion is to be eradicated.

Recently there have been a number of independent reports that the clinical presentation of actinomycosis may be more varied. *Actinomyces* spp. have been isolated in high numbers from acute soft tissue abscesses of the face. The large numbers of *Actinomyces* in these lesions suggest that they are involved in the aetiology of the infection, but their role is still unclear. It is possible that actinomycosis may also result from acute infections in which *Actinomyces* spp. gradually become the predominant micro-organism and the lesion becomes chronic.

Approximately 90% of all actinomycotic lesions occur in the face and neck region. The remaining 10% are predominantly abdominal and sometimes, but rarely, disseminated throughout the body. Disseminated infections can occur in medically-compromised patients and can be fatal if abscess formation occurs in vital organs (e.g. the brain). The abdominal lesions also include actinomycotic lesions associated with intrauterine devices, principally the birth control coil. Infections in this area can lead to abscesses in the fallopian tubes or the ovary, which can result in sterility.

Actinomycosis is a true opportunistic infection; the major sites where the organisms are found in a commensal state are the gingival crevice and tonsillar crypts. The presence of *Actinomyces* spp. in the uterine cervix associated with some contraceptive coils illustrates that the microbial ecology of a site can be drastically changed if the environment is altered.

Actinomyces israelii is still the most frequent isolate in actinomycosis, being recovered from over 80% of all lesions. *A. naeslundii*, *A. bovis* and *A. viscosus* are the principal isolates in the remaining 20% of major lesions. Other bacteria are also frequently present, and include *A. actinomycetemcomitans*, *Haemophilus* spp., *Propionibacterium* spp., 'Bacteroides' spp. and other obligate anaerobes.

The treatment of actinomycosis is by thorough surgical exploration of the lesion to drain the pus. Penicillin was the antibiotic of choice but, in order to get high therapeutic doses in the patient, intramuscular administration for at least six weeks was prescribed. To avoid this traumatic treatment oral penicillin derivatives such as amoxycillin, which have good absorption, are now recommended.

SALIVARY GLAND INFECTIONS

Sialadenitis

In health, the salivary glands are seldom infected. If, however, their function is impaired then infections can result. Bacterial infections

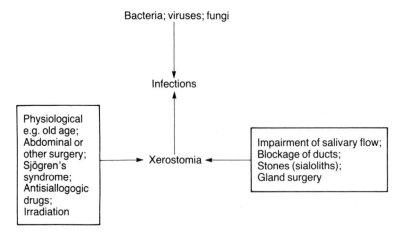

Figure 8.3 Predisposing factors to salivary gland infections.

usually result from micro-organisms entering the ducts or glands when the flow of saliva is interrupted. Figure 8.3 shows the conditions that predispose to salivary gland infections, all of which usually cause xerostomia (dry mouth). By far the most commonly affected salivary gland is the parotid. The infecting micro-organism can be viral or bacterial, and rarely fungal. The most common viral infection of the salivary glands is mumps.

Bacteria account for the majority of salivary gland infections, but the usual causative micro-organisms are not well defined. The reason for this is that salivary glands are difficult to sample without contamination from the resident oral microflora. The micro-organisms usually recovered are the oral streptococci, *Staph. aureus* and haemophili. A recent extensive study has shown that in some lesions '*Bacteroides*' spp. may be implicated.

Salivary gland infections believed to be bacterial are usually treated with a penicillin derivative. Unfortunately, the possibility of penicillinase-producing staphylococci cannot usually be eliminated when empirical treatment is being considered. Therefore, antibiotics such as flucloxacillin are usually the drug of choice. Antibiotic treatment can be supplemented with surgery to drain any pus.

The aetiology of salivary gland infections is complex and as yet not fully understood. Figure 8.3 illustrates some of the predisposing factors to infection; infection may occur once salivary flow is impaired and xerostomia results. A common cause of xerostomia is dehydration and this may occur particularly following general surgery where the patient fails to drink sufficient fluid. In many cases the cause of the loss of salivary flow may not be discovered and the use of sialogogues (salivary flow stimulants) may help to eradicate infection.

Halitosis

Halitosis is literally bad breath. It is caused by specific diseases or lack of oral hygiene. In particular, acute necrotizing ulcerative gingivitis (ANUG) (as described in Chapter 7), and other periodontal diseases may cause this condition. In recent years the cause of halitosis has been ascribed principally to the release of volatile sulphur compounds by the oral microflora. The two sulphur compounds that are most evident in halitosis are hydrogen sulphide and methyl mercaptan. Hydrogen sulphide is principally generated by the action of L-cysteine dehydro-sulphatase on L-cysteine. Methyl mercaptan is produced by the oxidation of L-methionine. The micro-organisms most frequently involved in the production of hydrogen sulphide and methyl mercaptan are peptostreptococci, *Porphyromonas* spp. and *Prevotella* spp. The incidence of halitosis increases with the deepening of pocket depth in the periodontal tissues. The deepening of pocket depth is accompanied by an increase in the flow of GCF. Amongst the constituents of GCF are serum proteins; these are broken down into peptides and eventually to constituent amino acids by plaque bacteria. The serum proteins are thus a source of L-cysteine and L-methionine for the production of halitosis. Treatment of the periodontal disease usually results in resolution of halitosis.

INFECTIVE ENDOCARDITIS AND BACTERAEMIA

The tooth-tissue interface is a unique site in the body which if diseased or damaged may offer relatively easy access to the blood stream for potentially pathogenic micro-organisms. The ingress of such micro-organisms is probably frequent but transient, the body's defence mechanisms eliminating them rapidly. Most dental procedures such as scaling, extractions, periodontal surgery and minor oral surgery produce bacteraemias, as can chewing or tooth-brushing. Bacteraemias can also occur in patients without teeth (edentulous), probably through minor breaks in the oral tissues.

Although most bacteraemias are of no consequence, in patients with cardiac disease, endocarditis can result. Table 8.3 lists conditions that predispose to infective endocarditis. All of these conditions alter the blood flow in the heart, particularly in the ventricles. The blood flow is therefore not uniform and 'eddy formation' occurs. 'Eddy formation' involves a swirling with a concomitant slowing at the edges of the flow, near the ventricular walls. Slow flow of blood encourages the formation of thrombi (clots) on the heart valves, particularly on the mitral valve at the junction of the left atrium and ventricle. These sterile thrombi, or vegetations, can be colonized as a result of a bacteraemia, and infections of the lining of the heart can result; this is called infective endocarditis. Such infections are serious, as they can result in thrombi breaking off and blocking the blood supply to a vital organ. They can severely impair the heart (failure) or they can cause a bacteraemia and an infection elsewhere in the body. Patients with these risk factors are given prophylactic antibiotic cover for most dental procedures.

Table 8.3 Valvular diseases of the heart that may cause a patient to be 'at risk' from endocarditis

Previous episodes of infective endocarditis
Prosthetic heart valves
Rheumatic fever
Aortic valvular disease
Mitral insufficiency
Ventricular septal defect
Patent ductus arteriosus
Valvular prolapses
Valvular stenosis
Degenerative aortic valve disease

Infective endocarditis is a serious life-threatening condition with a 30% mortality rate. It is difficult to diagnose and remains a problem to treat, despite modern antibiotics. It was formerly called subacute bacterial endocarditis, but this is an imprecise term for this condition as it is not subacute, and often not bacterial. If a patient recovers from infective endocarditis, then there is approximately a 40% probability that it will re-occur. The mortality rate for a recurrent episode of infective endocarditis is greater than 30%.

The micro-organisms that have been implicated in the pathogenesis of infective endocarditis are listed in Table 8.4. It can be seen that the oral streptococci are the predominant species isolated from infective endocarditis. The *S. oralis*-group are the oral streptococci most frequently found accounting for over 50% of isolations from the heart in infective endocarditis; of the remainder, the mutans streptococci (11%) and *S. salivarius* (10%) are the next most frequently isolated.

Table 8.4 Micro-organisms isolated from infective endocarditis

Micro-organism	%
Oral streptococci (e.g. *S. sanguis*, *S. oralis*, mutans streptococci)	36
Enterococci	14
Other non-Lancefield group streptococci	12
Staph. aureus	20
Non-coagulase staphylococci	10
Actinobacillus actinomycetemcomitans	3
Coxiella burnettii	3
Candida spp.	1
Other micro-organisms	1

A consideration of Table 8.4 shows that no single antibiotic would be effective against the entire range of micro-organisms isolated from infective endocarditis. This makes the choice of an antibiotic difficult to prevent infective endocarditis and consequently a compromise has to be effected. Originally penicillin was the drug of choice, but this has been superseded by orally-administered amoxycillin, which is well absorbed and has an appropriate spectrum of activity. Amoxycillin as a single 3 g dose has been shown to attain high serum concentrations which remain elevated for 7–8 hours after administration. Patients 'at risk' who are allergic to penicillin were given erythromycin derivatives, usually the stearate or the ethyl succinate. The doses of erythromycin required to attain good serum levels are high and this antibiotic has static rather than bactericidal properties. Erythromycin has been super-seded for prophylaxis by the use of clindamycin, which is bactericidal and well absorbed.

The majority of oral streptococci that cause infective endocarditis are in the *S. oralis*-group. These micro-organisms have been extensively studied (in a variety of animal models of infective endocarditis) to determine how they infect and how prophylaxis works. Much attention has been directed to the mode of attachment of members of the *S. oralis*-group to thrombi. As *S. sanguis* can catalyse the synthesis of extracellular glucan and fructan polymers (Chapter 4), it was suggested that these potential adhesins may aid attachment of *S. sanguis* to the thrombi. There is, however, no sucrose in blood (the substrate for polymer production) and the formation of glucan or fructan is unlikely to occur during bacteraemia. The duration of bacteraemias is also short (about 20 minutes at most) and the catalysis of sufficient polymer to mediate adhesion is unlikely to occur in this time. An alternative explanation with some experimental evidence to support it, is that *S. sanguis* possesses specific adhesins which allow it to attach to thrombi. Some of these putative adhesins have been identified and are polypeptides in the cell walls of *S. sanguis*. The precise method of attachment of *S. sanguis* to thrombi still remains unclear and further work is necessary to elucidate the attachment process.

THE USE AND MISUSE OF ANTIMICROBIAL AGENTS IN THE CONTROL OF MICRO-ORGANISMS

In 1983 the Editor of the *Lancet* wrote 'Before putting pen to paper, a good prescriber gives thought to the likely action of a drug. Where antimicrobial agents are concerned, there are two parties to be borne in mind besides the patient – the microbes and the environment.' There is no doubt that the widespread profligate, and in many cases, inappropriate use of antimicrobial agents in clinical practice is having a profound effect on micro-organisms. The treatment of oral infections with antibiotics is, in many cases, adding to this problem.

In recent years the value of an intact stable resident oral microflora has been appreciated as an integral part of the body's defences. The presence of an intact autochthonous oral microflora prevents colonization by exogenous pathogens or overgrowth by endogenous pathogens. This property is a vital component of the host's defences. Broad spectrum antibiotics destroy the integrity of the oral microflora and the colonization resistance by encouraging the selection and overgrowth of drug-resistant organisms.

If antimicrobials need to be prescribed for oral infections they should be used for the minimum period to obtain a cure. In some surgical textbooks there is still a common misconception that antimicrobials should always be taken for a set time, usually five days. It has been suggested, fallaciously, that failure to complete a course of antimicrobials could induce microbial resistance. Such an assertion is scientifically incorrect as drug-resistance is selected, not induced by antimicrobials. The minimal period of use of antimicrobials is necessary if colonization resistance is to be preserved.

The selection of resistance to antimicrobial agents has been reported in oral micro-organisms. Estimates of the incidence of penicillin-resistant oral streptococci have varied from 10% to 42% in volunteer studies. Resistance to other penicillin derivatives, such as amoxycillin, has been estimated as between 5–17% of streptococci, but is not so high in incidence as penicillin resistance. Some oral streptococci are resistant to the macrolides, and in particular erythromycin; these micro-organisms are often also resistant to lincosamides.

The existence of micro-organisms resistant to a single antibiotic is a worrying prospect, but multiple resistant oral bacteria have been reported. In the oral cavity infections due to multiple resistant strains of pneumococci, haemophili, *Neisseria*, *Branhamella* and *Staphylococcus* spp. have now been described. In general, the properties of antibiotic-resistant oral bacteria have not been studied in detail.

The emergence of resistant strains should reinforce the need for the cautionary use of antimicrobial agents. If antibiotics are used appropriately, by the correct route and for the minimum duration, then resistant micro-organisms will not be selected. Sadly, the inappropriate use of prophylactic and therapeutic antibiotics makes this unlikely.

SUMMARY

The potential exists in the oral cavity for the endogenous microflora to cause opportunistic infections if homeostasis is lost. These infections result in abscesses of teeth which may involve bone. If oral micro-organisms enter the blood stream, then infective endocarditis can result which can be fatal. A number of the operative procedures that are done in the oral cavity may also result in infection; usually this originates from the host's own microflora. The extensive use, and misuse, of antibiotics to prevent oral infections results in the selection of resistant organisms which may present problems for future therapy of infectious lesions.

FURTHER READING

British Society of Antimicrobial Chemotherapy (1990) The antibiotic prophylaxis of infective endocarditis. *Lancet*, **i**, 88–9.

Lewis, M. A. O., MacFarlane, T. W. and McGowan, D. A. (1990) A microbiological and clinical review of the acute dentoalveolar abscess. *British Journal of Oral and Maxillofacial Surgery*, **28**, 359–60.

Topazian, R. G. and Goldberg, M. H. (1987) In *Oral and Maxillofacial Infections*, (2nd ed.) (ed. W. B. Saunders), Philadelphia.

Samuels, R. H. A. and Martin, M. V. (1988) A clinical and microbiological study of actinomycetes in oral and cervicofacial lesions. *British Journal of Oral and Maxillofacial Surgery*, **20**, 458–63.

Stenhouse, D., MacDonald, D. G. and MacFarlane, T. W. (1975) Cevicofacial and intra-oral actinomycosis: a five year retrospective study. *British Journal of Oral Surgery*, **13**, 172–6.

9 Yeasts and viral infections

Yeasts and viruses are often responsible for oral infections. These infections have been the subject of intense research over recent years as they are often present in especially vulnerable patients, such as in the medically compromised.

CANDIDOSIS

A number of fungi, both perfect and imperfect, are known to infect or inhabit the oral cavity (Chapter 3). Amongst the perfect fungi (perfect = reproduce by sexual means) that cause infection are *Aspergillus* and *Mucor* spp. but these are rare. The commonest cause of oral fungal infections is the yeasts, and in particular *Candida* spp.

Candida spp. are common members of the oral microflora; the oral carriage rate is estimated at 40–60% of the population. The highest numbers of *Candida* spp. are thought to be present on the dorsum of the tongue. The exact division between candidal carriage and infection is often difficult to define. Attempts have been made to quantify the numbers of yeasts present in a specific area of oral mucosa using imprint culture techniques. This involves placing a square of foam soaked in culture medium on the mucosa, transferring it to a suitable medium and culturing it. The yeasts transferred by the foam grow into colonies and their number for a unit area can be calculated. Such techniques can be useful for detecting changes in numbers of yeasts present, but must be interpreted with caution. Healthy individuals may carry large numbers of yeasts, but have no signs, symptoms or pathological evidence of infection. Conversely, a small number of yeasts present in a medically-compromised individual may signify extensive infection. Thus, it is not just numbers of yeasts present at a site that is important, but their effect on the oral tissues.

The most commonly isolated *Candida* spp. from oral infections are *Candida albicans*, *C. tropicalis*, *C. glabrata* (formerly *Torulopsis glabrata*), *C. parapsilosis* and *C. guilliermondii*. One other yeast may be present, *Can-*

dida stellatoidea, but this is often difficult to distinguish from *C. albicans*. Most oral infections are thought to be caused by *C. albicans* and are opportunistic in nature, although other *Candida* spp. may also be present. There is some evidence that *C. tropicalis* and *C. parapsilosis* may be able to infect in synergy with *C. albicans*, but the precise mechanisms are still not understood.

Considerable debate still exists whether *Candida* infections should be called candid*osis* or candid*iasis*. In pathology the suffix ' -iasis' is reserved for parasitic infections. The term candidiasis is therefore imprecise. The term candidosis is correct, as the suffix ' -osis' reflects the fungal aetiology of this lesion and this will be the term used throughout this book.

C. albicans is often described as a dimorphic fungus in that it exists in blastospore and mycelial forms. When put onto certain specialized media (e.g. corn meal agar), small highly refractile spores, called chlamydospores, are formed. The exact function of chlamydospores is still unknown and they are not found in oral candidosis. There is still considerable debate about the role of blastospore and mycelial forms of *C. albicans* and the transition between these two forms. It has been suggested consistently that the transformation from a blastospore (yeast) to a mycelium, often called the Y-M transformation, is synonymous with a change from a commensal to pathogenic state. There is no doubt that increased numbers of mycelia are found in candidosis; however, mycelia can sometimes be found in healthy mouths. Mycelial-deficient *C. albicans* variants have been tested for their ability to produce candidosis and no clear answer has resulted. In some test systems the presence of the mycelial form is essential, whilst in others it may not be important. In general, however, the yeast mycelial transformation may be associated with candidosis.

The mycelia can in some conditions penetrate the oral tissues, but curiously they only penetrate the two superficial layers of epithelium, the keratin and granular layers, but never the whole thickness. When seen in biopsied tissues, the mycelia also appear only to penetrate roughly at right angles. The reason for this unusual mode of invasion is not clear. It has been suggested that the mycelia penetrate to obtain nutrients and to prevent displacement by desquamated oral epithelial cells. At present, although both explanations are plausible and have some scientific support, there is no definitive reason for these phenomena.

The virulence factors of *C. albicans* have been extensively studied and are listed in Table 9.1. There is a good deal of evidence that *C. albicans* adheres specifically to oral epithelial cells and to acrylic. This, together with mycelial penetration, gives a definite ecological advantage to the yeast. Once the yeast is attached to the oral epithelial surface, some strains may resist phagocytosis by either neutrophils or macrophages by the production of inhibitory acidic peptides. C. albicans can also bind components of complement to their surface and thus inactivate the complement cascade. C. albicans produces a whole variety of extracellular enzymes, in particular proteases and lipases. The presence of proteases in strains of C. albicans appears to enhance pathogenicity, but the role of the lipases is not as yet defined. The extracellular enzymes together with extracellular toxin-like glycoproteins and acid metabolites can also contribute to inhibition of phagocytosis, complement and the immune system. C. albicans can also form nitrosamines which may be important in chronic hyperplastic candidosis.

Table 9.1 Virulence factors of *Candida albicans*

Mechanisms	Molecular factors
Adherence	Extracellular enzymes:
Dimorphism	proteases
Interference with:	lipases
phagocytosis	Toxins
immune defences	Nitrosamines
complement	Acidic metabolites
Synergism with:	
bacteria and other	
yeast	

Oral candidosis is often described as a 'disease of the diseased'. This is because it is usually a secondary opportunistic infection to some other local or systemic predisposing disturbance. The predisposing factors that lead to oral candidosis are almost legion; some of the major factors are listed in Table 9.2.

Clinically, oral candidosis may be subdivided into:

Acute forms: pseudomembraneous candidosis; atrophic candidosis
Chronic forms: atrophic candidosis; hyperplastic candidosis; mucocutaneous candidosis
Other miscellaneous forms of candidosis

Table 9.2 Predisposing factors for oral candidosis

Local	Physiological
Trauma	Infancy
Occlusion	Old age
Maceration	
Saliva	**Hormonal states**
Xerostomia	Diabetes
Sjögren's syndrome	Hypothyroidism
Radiotherapy	Hyperparathyroidism
Cytotoxic therapy	Hypoadrenocorticism
Diet	**Nutrition**
High carbohydrate	Hypovitaminosis
	Iron deficiency
	Malnutrition

Acute pseudomembraneous candidosis (thrush)

Acute pseudomembraneous candidosis is a disease of newly-born children, old debilitated persons or those who are medically compromised. In newly-born children the source of the infection is usually the birth canal during parturition. The lack of microflora in the oral cavity at birth allows the *Candida* to flourish and to cause an infection. The cause of acute pseudomembraneous candidosis in elderly people is less well defined. It may be due to malnourishment, debilitation, xerostomia or impairment or atrophy of host defence systems. Thrush is also often present in medically-compromised patients, particularly in those infected with the Human Immunodeficiency Virus (HIV). In patients who are HIV-antibody positive the presence of acute pseudomembraneous candidosis is a bad sign, as it means that the host immune system is severely affected.

Acute pseudomembraneous candidosis presents as white plaques variable in size. Histologically these white plaques consist of dead mucosal cells and hyphal elements. If the pseudomembranes are removed then a raw bleeding area is left underneath. This form of candidosis is usually satisfactorily resolved by the use of topical antifungal agents, e.g. nystatin, miconazole. In AIDS patients the treatment is not so simple, as the condition progresses to involve the tonsils and the back of the throat. From here it can progress to affect the oesophagus or the trachea and the lungs. Thus, AIDS patients are usually given systemic antifungal agents such as fluconazole and ketoconazole.

Acute atrophic candidosis

Acute atrophic candidosis is caused by the suppression of the oral bacterial microflora by broad spectrum antibiotics, typically tetracyclines. There is a concomitant overgrowth by the oral fungi, in particular, C. albicans. The mucosa of the tongue and cheeks becomes thin, inflamed and atrophic in appearance. The condition is resolved by topical antifungal agents and cessation of the broad spectrum antibiotic.

Chronic atrophic candidosis

This condition is often called denture sore mouth, a misnomer as the patient is often completely unaware of it. Clinically it presents as a red, swollen, inflamed mucosa corresponding to the limits of the upper denture; it is seldom seen under lower dentures. Chronic atrophic candidosis is perhaps the commonest presentation of oral fungal infection, probably affecting 50 to 60% of all denture wearers. The principal yeast isolated is C. albicans, but others include C. tropicalis, C. stellatoides, C. parapsilosis, C. guilliermondii and, occasionally, Rhodotorula and Torulopsis spp. The principal site for yeast colonization is the denture itself and numerous hyphal forms can be seen in smears taken from the surface of affected cases. Curiously, at least four authors have now reported concomitant growth of Klebsiella spp. in cases of chronic atrophic candidosis, though the significance of this finding is unknown. Chronic atrophic candidosis is easily treated by scrupulous attention to denture hygiene, topical antifungal treatment and most importantly, simply leaving the dentures out of the mouth at night. Unfortunately, many denture wearers are reluctant to leave their prosthesis out at night.

Chronic hyperplastic candidosis

This curious condition is first seen as a white patch (leukoplakin) intra-orally, usually at the angles of the cheeks. The white patches cannot be rubbed off and if surgically removed are found to consist of grossly thickened epithelium penetrated by hyphal elements of C. albicans; the hyphal elements tend to be sparse. This condition is important in that 5–11% of all these lesions may become cancerous. It is an important but, as yet, unresolved question as to whether C. albicans is responsible for the cancerous change or just superinfects thickened tissue. At present the evidence, albeit indirect, is now strongly pointing to the fact that C. albicans is responsible for the cancerous change. If this lesion is treated with antifungal agents topically, systemically or both, then the lesions may resolve. It has also been shown that some strains of *C. albicans* can produce nitrosamines from saliva; nitrosamines are known carcinogens.

Chronic mucocutaneous candidosis

This is a rare condition which shows a predilection for young children or elderly males. It commonly affects the oral cavity first where thick granulomatous plaques form that are densely infiltrated with *C. albicans*. The granulomatous plaques may spread to involve the face, abdomen and nails where their removal or ulceration can lead to superinfection by bacteria such as pseudomonads, staphylococci or *Proteus* spp. Super-infection in cases of chronic mucocutaneous candidosis can be life-threatening. Recently the use of the systemic imidazole antifungal agent ketoconazole has revolutionized the treatment of this rare condition with many complete cures being reported.

Angular cheilitis

Angular cheilitis (perlèche) is an eroded condition of the angles of the mouth, particularly in skin folds. It can occur at any age, but it is often associated with denture wearers suffering from a concomitant chronic atrophic candidosis. Resolution of the chronic atrophic candidosis by topical antifungal treatment to the palate and angles of the mouth often results in its cure. In some patients, particularly the elderly, it is associated with a deficiency of folic acid and replacement therapy often results in its complete resolution. Many authorities have associated angular cheilitis with loss of facial height and formation of skin creases due to inadequate dentures. However, angular cheilitis does not always occur in edentulous patients who do not wear prostheses and curiously it is often seen in young patients with a natural dentition.

Angular cheilitis is virtually always associated with an overgrowth of *C. albicans* at the angles of the mouth. In addition, many bacteria are often found in large numbers, e.g. *Staphylococcus aureus*, anaerobic cocci and *S. pyogenes*. It is usually resolved by the use of antifungal agents with some bacteriostatic properties, e.g. miconazole; this is, of course, provided that there is no underlying deficiency state. Angular cheilitis is thus another example of an opportunistic infection.

Other miscellaneous forms of candidosis

Two other forms of candidosis have been described. These are cheilo-candidosis and juvenile juxtavermillion candidosis. They affect the lips and are probably either primary or secondary *Candida* infections.

OTHER FUNGAL INFECTIONS

Infections with perfect yeasts such as *Aspergillus*, *Geotrichium* or *Mucor* spp. do occur in the oral cavity, but they are rare. Their presenting features are varied and are usually secondary to trauma or extreme debilitation. Such infections are diagnosed from their histological appearance and are treated systemically. Increasingly these oral infections are being seen in AIDS patients.

THE HEPATITIS VIRUSES

No textbook of oral microbiology would be complete without some discussion of the hepatitis viruses. Strictly, the hepatitis viruses do not cause oral infections, but they are important in cross infection control (Chapter 10) and can be transferred in some cases by saliva and blood. Since the oral microbiologist is continually sampling areas where the hepatitis viruses occur, a thorough knowledge of them is required. Hepatitis viruses cause inflammation and damage to the liver. Four viruses are in this group: A, B, C and D. The essential features of these viruses are shown in Table 9.3. Hepatitis A is of little significance to oral microbiology, except that its mode of transmission is via the oral cavity. It is spread as a result of poor general hygiene or infected water supplies and causes no chronic liver disease.

Table 9.3 A comparison of the hepatitis viruses

Virus	Type of nucleic acid	Incubation period	Carrier state	Associated with prolonged chronic liver disease
Hepatitis A	RNA	Up to 40 days	No	No
Hepatitis B	DNA	2–6 months	Yes	Yes
Hepatitis C	Unknown	Unknown	Parenteral form only	Parenteral form only
Hepatitis D	RNA	Up to 40 days	Yes	Yes

In contrast, hepatitis B can cause chronic liver disease and has a significant mortality rate. It is transmitted principally by blood but is also secreted actively into the saliva. It is a DNA virus and exists in three forms, which are the filamentous, round and Dane particles. Only the Dane particle is the complete virus capable of replicating. The filamentous and round particles are the viral coat with no nucleus or associated enzymes. The presence of these particles is indicative of the fact that the virus does not always replicate successfully and abortive forms occur. The virus replicates in liver cells (hepatocytes) and the viral core, which contains a double strand of DNA, a DNA polymerase and another antigen (the HBe antigen) is assembled in the nucleus of the hepatocyte. The viral envelope is acquired from the hepatocyte cytoplasm as the virus particles are released. The released virus particles circulate in the blood stream, but it is thought that only the Dane particle is infectious. If the virus particles are transmitted by blood or bodily secretions then at least three consequences may ensue. The first is that the infected host may overcome the infection either because the viral dose is not sufficient to cause disease or there is natural immunity to the virus. The second possible consequence of infection is that the infected host may become subclinically infected and hence be an **asymptomatic carrier** of the disease; this state may persist for several months. The third consequence of viral infection is that the patient may become clinically ill, the first signs and symptoms resemble influenza, and jaundice is a rapid sequel. In the majority of cases the patient then recovers, the convalescent period often taking a few months. During the recovery period, and often beyond, the patient is still a carrier of the disease. Approximately 5% of all people who contract the disease have severe symptoms and prolonged life-threatening illnesses which may result in death by loss of liver function, or from an opportunistic secondary bacterial infection. There is at present no successful treatment and hence, only palliation of the clinical course of the disease can be used. Patients who survive severe hepatitis B may take months or years of convalescence to recover, during which time they remain asymptomatic carriers of the disease. Unfortunately, a minority of these patients subsequently develop primary hepatocellular carcinoma for which the prognosis is poor.

Thus, it can be seen that the course of hepatitis B is variable and potentially life-threatening. Since there is no known cure for the disease the best course for its eradication is prevention by recognition and elimination of carriers. It has been estimated that there are between 120 and 175 million asymptomatic carriers of the disease worldwide. The prevalence of carriers varies; in America and Northern Europe the carrier rate is less than 0.1% of the population, in Central Europe it is 5%, whilst in the Mediterranean countries and some parts of Africa and

Asia as many as 20% of the population may be carriers. Since the disease is transmitted by bodily fluids, a large proportion of drug addicts and those who have received unscreened blood products may carry the disease. Hepatitis B has also been shown to be a sexually-transmitted disease with a predilection for homosexuals. One other group of potential carriers deserves a mention – this is institutionalized patients suffering from Down's syndrome (mongolism) who for, as yet, unexplained reasons have very high carriage rates.

With such a large and diverse group of carriers of the disease the potential for transmission is high. Thus, it is important that carriers are identified if they are to undergo any surgical or microbiological procedure. There are now a variety of very sensitive serological tests available for the recognition of potential carriers of hepatitis B, which can be undertaken by specialist laboratories; these tests are not, however, suitable for use in routine dental practice.

The serological tests for hepatitis B are based on the detection of two circulating antigens in affected patients. These are the surface and e antigens. The surface antigen was first discovered quite accidentally, by Blumberg, in an Australian aborigine who was a carrier of the disease (hence the origin of the term, Australia antigen). Blumberg pioneered the serological detection of hepatitis B and for this he was awarded the Nobel prize. The presence of surface antigen in a patient's serum shows the patient may have had active disease or, alternatively may be an asymptomatic carrier. Thus, nowadays more reliance is placed on determining whether the e antigen is present, or the presence of antibodies against e. This shows that the whole virus is present. The presence of the hepatitis B e antigen is indicative of clinical disease and of the possibility that transmission may occur. Recently another serological marker has been detected, the so-called Delta agent; this appears to be limited to cases of hepatitis in which liver cancer subsequently develops. The development of early detection of the Delta agent may help to screen for patients likely to develop liver cancer.

One approach to prevention is the recognition of carriers, but if the disease is to be eradicated on a world-wide scale then a vaccine is essential. Two vaccines are currently available. The first is manufactured by the purification of the viral surface antigen (HBsAg) from the sera of patients with hepatitis B. This vaccine is effective (> 95% protection), but is expensive. The serum-derived vaccine has been superseded by a genetically engineered product. The genes coding for the surface antigen have been cloned and genetically engineered into the yeast *Saccharomyces cerevisiae*. The inserted genes are functional in this yeast and produce hepatitis B surface antigen. This is purified and used for vaccination.

Non-A, non-B hepatitis has been renamed as hepatitis C. This is, in fact, a group of viruses. Little detailed information is available about the viruses causing hepatitis C, but the mode of transmission has been identified for two of the group. Epidemic hepatitis C is spread as a result of poor general hygiene or infected water supplies. It causes a self limiting type of jaundice and no chronic liver disease. Its pattern of infection is essentially similar to hepatitis A. In contrast, the parenterally-transmitted form of hepatitis C can cause a serious illness and death. It is transmitted usually by blood but salivary transmission has been described in experimental animals. It causes chronic liver damage in 25% of persons affected by the disease. The mortality rate is higher than hepatitis B but is not precisely evaluated yet. Serological tests for both forms of hepatitis C are now available and thus precise epidemiology on these viruses will be possible. At present there is neither a vaccine nor treatment available for hepatitis C.

Hepatitis D or delta hepatitis is a defective RNA virus which needs the presence of pre-existing hepatitis B infection in order to establish itself. It causes chronic active hepatitis and is prevalent in parts of the Middle East and Africa. It is particularly associated with intravenous drug abuse and is spread by blood-to-blood contact. Vaccination to prevent hepatitis B will also stop hepatitis D infection.

HUMAN IMMUNODEFICIENCY VIRUS

HIV infection affects T 'helper' lymphocytes. Once infected, the host's immune system may be so impaired that opportunistic infections and other changes can occur in the oral cavity. The range of changes possible is large and detailed in Table 9.4. Most of the infections occur as a result of impaired or non-existent immune responses to the micro-organisms. There has not yet been a detailed description of the natural history of the changes in the oral microflora of patients infected with HIV. What is known is that profound changes occur and that non-resident Gram-negative bacteria are able to colonize the mucous membranes. These bacteria include *Pseudomonas*, *Klebsiella* and *Escherichia* spp. The exact effect of such colonization is not known, but it has been postulated that the endotoxins produced by these bacteria deleteriously affect the oral tissues and predispose them to further infection.

There is some speculation as to whether HIV is transmitted by saliva. There is a considerable amount of evidence principally generated from studies of families containing an HIV antibody-positive person that direct contact or saliva transmission is unlikely. The virus can be found occasionally in saliva, but it is known to be rapidly inactivated. Recently, however, HIV has been transmitted from a dentist to his patients,

Table 9.4 Oral lesions associated with HIV infection

Fungal infections
  Candidosis
    Pseudomembranous
    Erythematous
    Hyperplastic
    Angular cheilitis
  Histoplasmosis
  Cryptococcus neoformans
  Geotrichosis

Bacterial infections
  HIV-necrotizing gingivitis
  HIV-periodontitis
  Mycobacterial infections
  Klebsiella pneumoniae
  Enterobacterium cloacae
  Escherichia coli
  Sinusitis
  Exacerbation of apical
    periodontitis
  Submandibular cellulitis

Viral infections
  Herpetic stomatitis
  Cytomegalovirus
  Hairy leukoplakia
  Zoster
  Varicella

Papilloma virus lesions
  Verruca vulgaris
  Condyloma acuminatum
  Focal epithelial hyperplasia

Neoplasms
  Kaposi's sarcoma
  Squamous cell carcinoma
  Non-Hodgkin's lymphoma

Neurological disturbances
  Paraesthesia
  Facial palsy

Unknown aetiology
  Recurrent aphthous ulceration
  Progressive necrotizing
    ulceration
  Toxic epidermolysis
  Delayed wound healing
  Idiopathic thrombocytopenia
  Salivary gland enlargement
  Xerostomia
  HIV-embryopathy
  Submandibular
    lymphadenopathy
  Hyperpigmentation

possibly as a result of contaminated instruments. Saliva and blood were probably involved in this transmission.

The major risk groups for HIV are shown in Table 9.5. This group does not include the dental profession. The major route of transmission is blood contact and even this is not always successful (Chapter 10). HIV is not very infectious and a large number of particles are necessary

Table 9.5 Major groups at risk of HIV infection

Sexually-promiscuous homosexuals or bisexuals
Intravenous drug abusers
Recipients of unscreened blood products (e.g. haemophiliacs)
Sexual partners of the above
Children born to infected mothers

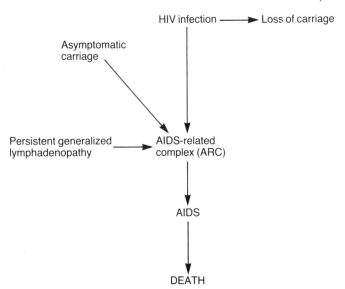

Figure 9.1 Possible consequences of HIV infection.

to cause infection. The possible consequences of HIV infection are shown in Figure 9.1.

HIV infection usually leads to the Acquired Immune Deficiency Syndrome (AIDS). There have been two well-documented cases of HIV infection being lost, but this is rare. The length of asymptomatic carriage is variable and may be as much as nine years. The effect on the immune system may be variable at first, but as more lymphocytes become affected it becomes more profound. The first sign of infection is usually general and persistent enlargement of lymph nodes; this is called persistent generalized lymphadenopathy (PGL). The appearance of some signs or symptoms of AIDS in infected individuals is called AIDS-related complex or ARC. In some patients this is a transitory and rapid stage before the full syndrome supersedes. The signs and symptoms of the syndrome are numerous and varied. The principal ones are weight loss, malaise, night sweats, PGL and opportunistic infections (especially of the lungs and mouth). Death is usually caused by infection.

The HIV viruses are RNA viruses. They are retroviruses which have an enzyme, reverse transcriptase, associated with their genome. Once inside a lymphocyte the RNA is transcribed into DNA and retroscribed back into viral RNA; this is done by the reverse transcriptase. Direct detection of the virus is difficult. Tests for carriage rely therefore on the production of antibody. In some individuals antibody may not be expressed for at least eleven months after infection, so false negative tests are possible.

HERPES

The herpes group of viruses includes *Herpes simplex* types I and II, varicella zoster, cytomegalovirus and Epstein Barr virus. By far the most common infection by this group of viruses is caused by *Herpes simplex* type I (HSV–1). HSV–1 is a large DNA virus which affects over 90% of the population. It is transmitted by direct contact or by saliva. Herpetic infections are either primary or secondary. The most common primary infection is herpetogingivostomatitis (Chapter 7). In the course of the infection the virus travels along sensory nerves to the trigeminal ganglion. Here the virus remains latent until reactivated. Various stimuli can cause reactivation and they include, common colds, sunlight (possibly by ultraviolet radiation), stress and menstruation. The virus returns along the sensory nerves to the mucous membranes and causes the lesion called 'cold sores'. Although most people are affected by the virus, only about 30% suffer from cold sores. Reactivation of the virus in the ganglia can result in the excretion of the virus in saliva without cold sore formation; the frequency of such excretion without lesions is still unknown. Cold sores occur usually at junctions between mucous membranes and skin. They heal spontaneously without leaving any scars. Nowadays the discomfort of cold sores can be ameliorated by the use of acyclovir, a guanine analogue. Acyclovir prevents viral replication and is best applied in the prodromal phase of cold sores. The prodromal phase is identified as the prickling or burning sensation that often precedes cold sores.

Varicella zoster produces crops of painful intraoral vesicles which are distributed down the course of a sensory nerve. The vesicles heal in 7–10 days with no scarring. This condition is often called shingles and is a reactivation of varicella zoster; the primary illness is chicken pox. About 15% of all shingles is intraoral and may lead to postherpetic neuralgia. This is a painful condition in which there is constant or intermittant pain along the course of the sensory nerve affected. The treatment for shingles is acyclovir and there is evidence that if this is used promptly then post-herpetic neuralgia can be prevented.

HAND, FOOT AND MOUTH DISEASE

Hand, foot and mouth disease causes small epidemics in schools, hospitals or other community institutions. It is caused by Coxsackie virus A16 or A4, A5, A9 or A10. The onset of symptoms is a mild fever with vesicles on the skin of hand or foot and on the oral mucous membranes. In the mouth the lesions can be on the gingivae, buccal mucosa or tongue and occasionally on the soft palate or tonsillar area. The lesions heal spontaneously within seven days and no treatment is required. While the lesions are present the virus can be transmitted.

OTHER VIRUSES

Other viruses can cause oral lesions, but they are rare. These viruses include other Coxsackie A2 infections, which are responsible for herpangina. Herpangina affects the pharynx principally or the tonsillar area and appears as a reddened painful area. Occasionally herpangina can affect the parotid gland but this is rare. Herpangina is self limiting, but can be very distressing for the patient; no definitive treatment is available.

Measles and mumps both can cause oral lesions. Measles produces the tiny, bluish-white spots with red areolae (Koplik's spots) on the mucous membranes near the molar teeth. Mumps affects the pharynx and commonly the parotid salivary glands which swell and become painful. Both mumps and vesicles resolve spontaneously in western populations, but in Africa both can cause substantial mortality, especially in children.

SUMMARY

Yeasts and viruses can cause a whole range of lesions in the oral cavity. Yeasts principally affect patients whose host defences are debilitated. The principal cause of yeast infections in the oral cavity is *Candida albicans*.

Viruses similarly cause a number of characteristic lesions (Table 9.6), the most common of which are 'cold sores'. The AIDS epidemic has produced a number of oral infections which are due to the compromised immune status of the host.

Table 9.6 Viruses that can cause lesions in the oral cavity

Virus	Oral lesion
Herpes simplex type 1	Cold sores
	Gingivostomatitis
Herpes simplex type 2	Oral lesions
Varicella zoster	Chickenpox
	Shingles
Hand, foot and mouth disease	Hand, foot and mouth lesions
Herpangina	Mouth lesions
Measles	Koplik's spots
Mumps	Parotid swelling
HIV	Various lesions

FURTHER READING

Oral Candidosis (1990) (eds. L. P. Samaranayake and T. W. MacFarlane) Wright, London.

Timbury, M. C. (1983) *Notes on Medical Virology*. Churchill Livingstone, Edinburgh.

Topazian, R. C. and Goldberg, M. H. (1987) *Oral and Maxillofacial Infections*, Saunder and Company, London.

Topley and Wilson's Principles of Bacteriology, Virology and Immunity (1990) 8th edn, Vol. 4. (Eds M. T. Parker and L. H. Collier) Edward Arnold, London.

10 Cross infection control

One of the results of the AIDS epidemic is that the interest of the dental profession in general, and of oral microbiologists in particular, has been focused on cross infection control. Prior to 1980 cross infection control had been given only perfunctory attention and many major textbooks of dental surgery practice at that time had suggested that 'kitchen cleanliness' was all that was required. Nowadays, strict surgical aseptic techniques are required, and anything less may have medico-legal consequences.

Cross infection usually results from the transfer of *exogenous* micro-organisms (organisms derived from outside of the host's oral cavity). Cross infection is distinct from *opportunistic* infections which arise from the host's own *endogenous* microflora. The sources of the endogenous and exogenous micro-organisms are shown in Table 10.1. The personnel at risk from cross infection are shown in Figure 10.1. It should be noted that all the staff involved in health care, as well as the patient, are at risk from cross infection.

Table 10.1 Sources of infectious agents

Exogenous	Endogenous
Atmosphere	Saliva
Dust/Dirt	Blood
Water	Gingival crevicular fluid (GCF)
Instruments	Skin
Materials/drugs	Faeces

The transmission of infection in dentistry is possible by four main routes. These are *contact, ingestion, inhalation* and *inoculation*. Contact may involve direct transfer from skin to skin, or from skin to mucous

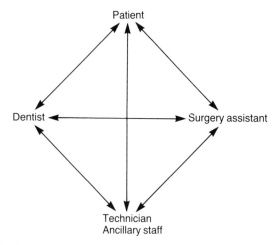

Figure 10.1 Pathways for the transmission of infection in dentistry.

membrane; in the latter case saliva may be involved. For example, direct contact can result in the transfer of herpes and staphylococci. Dental procedures can cause ingestion via swallowing, and result in the transfer of viruses such as hepatitis A. Inevitably, the use of coolants for such instruments as hand-piece turbines causes aerosol formation.

The evidence for the transfer of pathogens in dentally-generated aerosols is inconclusive, but theoretically micro-organisms such as *Legionella* spp., *Mycobacterium* spp. (e.g. *M. tuberculosis*) and influenza could be transferred. Inoculation is a proven cause of cross infection, and the transfer of micro-organisms such as hepatitis B and HIV by this route has been proved conclusively. A distinction must be made between cross infection and cross contamination. Cross infection is the transfer of micro-organisms that results in an infection. Cross-contamination is the transfer of micro-organisms to a host; it may or may not result in infection, although in practice, it does not usually result in infection. The reasons for this are usually that insufficient micro-organisms are transferred (i.e. below the infective dose), they fail to colonize, or they are eradicated by the host's defences. Perhaps the most potent protection the host has is the resident microflora itself, which can prevent colonization by a number of mechanisms (colonization resistance) (Chapter 4; Table 4.7). Other host defence mechanisms (Chapter 2) are also effective in preventing infection.

THE EVIDENCE FOR PROVEN CROSS INFECTION

The incidence of cross infection in dentistry is probably underestimated as usually the infectious disease is not related to a visit to the dentist. Some of the microbiological evidence that cross infection does occur in dentistry is summarized in Table 10.2. The strongest evidence is provided by the transmission of hepatitis B. Before vaccination against hepatitis B became universal amongst dentists, practitioners were infected. Since vaccination has been introduced the number of dental personnel infected with hepatitis B has fallen dramatically.

Table 10.2 The micro-organisms involved in proven cases of cross infection; their source and probable route of transmission are listed

Micro-organism	Source	Probable route of transmission	Result
Herpes simplex type 1	Dental hygienist	Hands	Oral herpes
Herpes simplex type 1	Patient	Saliva	Herpetic whitlow
Hepatitis B	Dentist	Non-sterile instruments	Hepatitis B
Hepatitis B	Dentist	Non-sterile needles	Hepatitis B
Hepatitis B	Patient	Saliva	Hepatitis B
HIV	Patient	Needlestick injury	HIV antibody positive dentist
HIV	Dentist	Contaminated instruments	Infected patients
Hand, foot and mouth disease	Dentists and patients	Bodily contact, secretions	Typical lesions of hand, foot and mouth disease
Staphylococcus aureus (multiple-resistant)	Dentist	Hands	Dental abscesses
Pseudomonas aeruginosa	Dental unit water supplies	Water coolant	Dental abscesses
Other bacteria	Dental procedures	Aerosols? Instruments	Occasional reports

Hepatitis B can remain viable outside the body, especially in blood, for periods of hours. It is therefore important that all instruments are thoroughly cleaned and sterilized. Hepatitis B has been transmitted by infected saliva and blood through instruments, especially needles.

The case for transmission of HIV is less clear. One dentist who had repeated needlestick injuries has become antibody-positive for HIV. The source of the HIV virus is not known but is probably an infected patient. In addition, one HIV antibody-positive dentist has infected at least three of his patients, probably by operating with instruments previously used on himself. In general, HIV is regarded as less infectious than Hepatitis B.

Herpes simplex type 1 infections have long been recognized as being caused by cross infection. There have been numerous reports of herpetic whitlows on dentists' hands, and these are a result of contact with infected lesions. This is hardly surprising as 98% of the population carry the virus. There is also a well documented case of herpes being transmitted to five patients by a dental hygienist, probably by direct contact. In addition, there is evidence that *Herpes simplex* type 1 can remain viable on record cards, especially if mixed with blood.

Hand, foot and mouth disease has been reported in epidemics affecting large institutions such as dental schools and hospitals. This is spread as a result of poor aseptic techniques; fortunately, the disease, though inconvenient, is usually self-limiting.

The evidence for bacterial cross infection is sparse. This is probably due to the fact that in normal patients the resident microflora is effective in preventing the establishment of exogenous bacteria by colonization resistance. In addition, many of the potential invading bacteria cannot specifically attach to mucous membranes. In medically compromised patients, these exogenous bacteria are able to cause infections due to the host's defences being impaired. Thus, Gram-negative bacteria from contaminated dental unit water supplies can cause problems. There has been a report of *Peudomonas aeruginosa* being spread from dental unit water supplies (DUWS) and causing minor oral infections in two medically compromised patients. DUWS do get contaminated as a result of back-siphonage from dental instruments such as hand-piece turbines; they do need periodic disinfection to remove this contamination. In theory *Legionella* could be spread by the coolants used in dentistry. The use of high vacuum aspiration during operative procedures usually prevents large droplets (which are necessary for transmission of *Legionella*) from leaving the mouth; thus, no case of legionellosis has been conclusively linked to dentistry. Certainly, *Legionella* antibodies are present in high titres in some dentists, but no cases of legionellosis have been reported from this exposure. *Staphylococcus aureus* has been linked with two oral abscesses; the transfer was probably by the dentist's hands.

It is perhaps surprising that the reported incidence of cross infection is so low. Summating all the evidence, both circumstantial and proven, there is no doubt that cross infection does occur as a result of dental procedures. The exact magnitude of cross infection in dentistry is unlikely to be known as surveys are difficult if not impossible to complete accurately. More importantly, the nature of dental aseptic practices have recently changed so radically that a baseline figure for incidence would be impossible to obtain. Fortunately, colonization resistance and the array of specific and innate host defences are highly effective in preventing colonization and infection.

VACCINATION

Vaccination of dentists and ancillary personnel is a reliable and important part of cross infection control. The vaccinations that are important are hepatitis B, tuberculosis, tetanus and poliomyelitis; women of childbearing age should also receive a Rubella vaccination. A summary of the vaccinations and their route of administration is shown in Table 10.3. The need for tetanus vaccination may not be immediately apparent, but this rare disease may be transmitted by spores in carious dentine. It is important to note that *all* personnel should be vaccinated; this includes ancillary and support staff.

Table 10.3 Vaccinations required by dental personnel

Vaccine	Route	Booster period
Hepatitis B	Intramuscular 3 injections: 0, 1, 6 months	3–5 years
Tuberculosis	Intramuscular	Retest every 5 years
Poliomyelitis	Oral	5 years
Tetanus	Intramuscular	5 years
Rubella	Intramuscular	> 15 years

MEDICAL HISTORY

A thorough comprehensive medical history is a pre-requisite before any form of dental treatment. It is helpful to know that a patient has a potentially transmissible disease and this has led some authorities to divide patients into 'high' and 'low' risk categories. Some regard such a division as being both impractical and dangerous. Many patients do not know they are infected (e.g. asymptomatic carriers), or are 'economical with the truth'. Thus, the safest policy is to assume that everyone is infected and to design one set of cross infection procedures safe for every patient. Thus, whilst the information that a patient is infected with, for example, HIV or hepatitis B, may be important in anticipating oral problems, it should not influence the standard and safe implementation of cross infection procedures.

THE RISK AREAS

When the AIDS epidemic first became apparent the paranoia and fear among the dental profession resulted in the rapid issue of cross infection guidelines. Many of these guidelines, designed with the knowledge of HIV available at the time, were over-cautious and impractical. As knowledge of HIV has improved more practical guidelines have been published. These revised guidelines take into account the real risks to dental personnel. It is now possible to define the risk areas to dental personnel and these are listed in Table 10.4.

Table 10.4 The risk areas for cross infection in dentistry and preventive measures

Risk area	Prevention
Hands	Gloves and handwashing
Needlestick injuries	Improved technique; safe disposal of 'sharps'
Instruments	Sterilization and disinfection
Eyes	Protective spectacles
Surfaces	Disinfectants/zones
Aerosols and splatter	High-vacuum aspiration/masks/ protective spectacles
Infected waste	Incineration

Hands

Unprotected hands present a real danger of cross infection. Most hands have minute cuts or abrasions on them and therefore provide the opportunity for blood-to-blood contact. The simple remedy for this is to wear protective gloves for all routine dental operations. This measure has been adopted by the majority of practitioners, but it has not been without problems. Many gloves have caused sweat-retention and other irritant dermatological reactions. In a minority of practititioners there have been hypersensitivity reactions. In general, unavoidable reactions to gloves are rare and proper handcare will prevent problems.

Handwashing is also an important part of cross infection control. The proper use of liquid soap and disinfectant has been shown to reduce the transient microflora on hands by nearly 80%; with repeated washing this increases to over 90%. Reduction of the transient microflora of the hand prevents the possibility of cross infection from dentist to patient. The use of a disinfectant beneath a glove also helps if the glove is accidentally punctured. Any micro-organisms that go through the glove to the bare hand will meet disinfectant, and may not be able to infect.

The re-use of gloves is a controversial topic. Many authorities maintain that non-sterile gloves cannot be reused. Ideally, gloves should be changed after every patient, but in practice they are often rewashed. Investigations of this problem have shown that careful rewashing of gloves for a limited number of times (about 4–6, with good quality gloves) is probably safe.

Needlestick injuries

Needlestick injuries are a serious problem in dentistry. The transmission of viruses, such as HIV, has been proven to occur through blood-to-blood contact. Needlestick injuries arise in dentistry usually through the resheathing of needles or by mishandling instruments. Dentistry is unusual in that it is probably the only healthcare profession with a genuine reason to resheath needles. The reason for resheathing is that it is often necessary to give local anaesthetic more than once during the course of a procedure; using a fresh needle for each injection is uneconomic and wasteful. Leaving a needle unsheathed is, of course, dangerous. Needles can in practice be resheathed safely. The simplest method is to insert the needles into the sheath holding only the barrel of the syringe. This method is often called the 'bayonet' technique. Keeping one hand on the barrel avoids the other hand being in proximity to the needle; the latter is the cause of most injuries. Once the sheath is on the needle, the instrument is held vertical and the sheath secured with the other hand. If the bayonet technique is not used then a whole variety of needleguards are commercially available. These devices allow the needle to be resheathed one-handed or protect the hand that holds the sheath.

The other type of needlestick injury occurs due to penetrating the epithelium by other sharp instruments used in dentistry. Although this injury can occur at any time, it usually occurs when dental surgery assistants are washing instruments prior to sterilization. This latter injury can be avoided by the use of heavy duty gloves.

The treatment of needlestick injuries has caused concern in recent years due to the fear of HIV transmission. A large number of needlestick injuries from HIV 'donors' have been documented. The risk of transmission appears to be about 0.1%. The immediate treatment of needlestick injuries is essential. The wound should be encouraged to bleed as copiously as possible. If the recipient has not received Hepatitis B vaccination then this should be commenced immediately. In some centres, if the 'donor' is hepatitis B positive then hyperimmune gammaglobulin is given; this is globulin from patients known to have had hepatitis B. In all cases, a sample of the recipient's serum should be stored for medico-legal reasons; the exact details of the injury should also be recorded. Many authorities recommend the prophylactic use of azothymidine for needlestick injuries, but its value in the prevention of HIV has not been fully proven.

Instruments

In general, all instruments should be sterilized. Nowadays there are only a minority of instruments that cannot be sterilized and these should be disinfected or be regarded as disposable. Three methods of sterilization are currently available and these are summarized in Table 10.5. Instrument sterilization involves four distinct processes; these are (1) pre-sterilization; (2) cleaning; (3) the sterilization process, and (4) aseptic storage. All instruments need thorough cleaning of all detritus prior to sterilization. This is important as micro-organisms can remain viable if 'insulated' by organic matter. In practice, the cleaning of intricate instruments can be facilitated by the use of ultrasonic baths followed by careful inspection. Handling of unsterilized instruments should be done with care and always with the protection of heavy duty gloves.

The two methods of choice for the sterilization process are the autoclave and the chemiclave. The autoclave uses the latent heat of steam under pressure. A variety of temperature, time and pressure combinations can be used, but 121°C at one bar for 15 minutes or 134°C at 2 bar for 4 minutes are the most usual combinations. In general, autoclaves used in dentistry are non-vacuum in operation; this means that the air is purged from the chamber by the incoming steam. Autoclaves have the advantage that they are reliable, cheap to run, and the sterilization cycle cannot be interfered with once started. The chemiclave employs chemical vapours at high temperatures (132°C) for 20 minutes

Table 10.5 A summary of the methods of instrument sterilization

Method	Principle	Advantages	Disadvantages
Autoclave	Steam under pressure	Running costs low Cycles cannot be interrupted	Apparatus is expensive; rusts dental instruments
Chemiclave	Various microbiocidal substances at high temperatures, usually in vapour form	Does not rust dental instruments Cycles cannot be interrupted	Long cycle time; vapours may be harmful; Machine needs careful maintenance Chemicals involved are expensive
Hot air ovens	Hot air	Very cheap to run	Long cycle; not reliable even when fitted with circulatory fans; damages instruments; sterilization cycles can be interrupted

to sterilize instruments. The great advantage of the chemiclave is that it does not rust instruments and therefore is particularly useful for orthodontics. The main disadvantage of the chemiclave is that the vapours it uses are toxic and are known carcinogens (e.g. ethylmethyl ketone and formaldehyde). It must therefore be regularly serviced to prevent toxic leaks. The other type of sterilization process is hot air. This method employs an oven, often with internal circulation fans. The sterilization cycle is long (180°C for 30 minutes or, 160°C for one hour), and this does not include the heating-up or cooling-down time. Hot air ovens can have their sterilization cycles interrupted and more importantly often do not reach or maintain the recommended temperatures. In addition, they do damage delicate instruments and are therefore no longer recommended for dental practice.

The last stage of sterilization is aseptic storage. In general instruments should be stored dry in impermeable containers. A whole variety of such containers are commercially available.

Eyes

Damage to the eyes can easily occur during dental procedures. The removal of a large amalgam filling with a rotary instrument can generate particulate matter that leaves the mouth at speeds in excess of 500 mph. High speed instruments can also generate droplets which can splatter and contaminate the eye. The simple use of protective spectacles protects the eyes from this danger.

Surfaces

Surfaces do become heavily contaminated during dental procedures usually by touch, placing contaminated instruments on them, and from aerosols and splatter. To help in the decontamination of such surfaces, the areas that are regularly touched or used for instruments are best restricted in use; this principle is known as zoning. The designated contaminated areas or zones are easily identified and can be thoroughly disinfected with suitable agents. A large number of commercially available disinfectants are now available and in some countries these are the subject of extensive regulations. A suitable disinfectant should be chosen from those recommended by the national regulatory body.

Aerosols and splatter

The dangers of cross infection through aerosols in dentistry is not proven. Droplets are generated but they are small and probably do not contain enough micro-organisms to cause an infection. In contrast, large droplets are occasionally ejected from the mouth and can cause eye infections; this can be prevented by the use of protective spectacles. In practice a good quality mask prevents aerosols and splatter from reaching the face. It must be emphasized, however, that such masks are not a good form of microbiological protection. Once wet, masks become permeable to micro-organisms.

Infected waste

Two types of infected waste are generated from dental practice; these are sharps and other materials. Sharps should be disposed of in rigid containers by incineration. Containers should never be filled more than three-quarters full before disposal to prevent accidental injury when inserting the sharps. Other clinical waste should be separated from sharps and put into stout bags. Ideally these should be incinerated, but in some countries they can be buried in deep refuse holes.

SUMMARY

The risk areas where cross infection can occur have now been identified in dentistry. They are hands, needlestick injuries, eyes, surfaces, aerosols and splatter, and infected waste. Procedures have now been devised to reduce the risks in all these areas to allow safe dental practice.

FURTHER READING

American Dental Association Council on Dental Therapeutics and Council on Prosthetic Services and Dental Laboratory Relations, Guidelines for Infection Control in the Dental Office and the Commercial Dental Laboratory (1985) *Journal of the American Dental Association*, **110**, 969–72.

Anonymous (1988) The control of cross infection in dentistry. *British Dental Journal*, **165**, 353–4.

Field, E. A. and King, C. M. (1990) Skin problems associated with routine wearing of protective gloves in dental practice. *British Dental Journal*, **169**, 281–5.

Martin, M. V. (1991) *Infection Control in the Dental Environment: Effective Procedures*, Martin Dunnitz, London.

Concluding remarks

Impetus for oral microbiological research has come from the finding of an association between dental plaque and two of the most prevalent infections known to man – caries and periodontal disease. Both infections result from the complex interaction of diet, the oral microflora, and the host. Consequently, an understanding of the ecology of the mouth is essential to determine the processes involved in the pathogenesis of these conditions.

The composition and metabolism of the resident oral microflora is dictated by the properties of the mouth as a microbial habitat. Several features contribute towards making the mouth unique in this respect. Teeth provide hard, non-shedding surfaces for colonization. This allows the accumulation of large masses of micro-organisms (predominantly bacteria) which together with their extracellular products and adsorbed salivary polymers is termed dental plaque. The mouth has also several distinct epithelial surfaces for colonization, such as lips, palate, cheek, gums and tongue. Saliva and gingival crevicular fluid (GCF) provide a continual supply of endogenous nutrients that can be utilized by the resident microflora.

Bacteria interact synergistically to metabolize these host macromolecules. Saliva and GCF also introduce components of the immune response (humoral and cell mediated) and innate host defences which help regulate the growth and metabolism of organisms in the mouth. The flow of saliva and GCF remove weakly-adherent cells, so that many organisms are restricted to sites offering protection from the environment, such as the fissures and approximal surfaces on teeth.

The biological properties of the mouth make it highly selective in terms of the micro-organisms that are able to colonize and multiply. Despite this selectivity, the resident oral microflora supports the growth of viruses, fungi, (occasionally protozoa), and a diverse range of bacteria including facultative and obligately anaerobic species. These micro-organisms are not evenly distributed around the mouth; the location of many species is dictated by the redox potential at a site, the provision of essential nutrients, and the strength and specificity of adherence to a surface. For example, obligate anaerobes are located mainly in dental plaque and on the tongue; *S. sanguis* and mutans streptococci have a preference for hard surfaces while *S. salivarius* is found in higher numbers on the oral mucosa.

Adherence involves specific interactions between adhesins on the microbial cell surface and receptors (ligands) on the host. Most research has been directed towards the mechanisms involved in the formation of dental plaque. Clean enamel is rapidly coated by a film of proteins and glycoproteins adsorbed from saliva. Initially, bacteria interact reversibly with this pellicle-coated enamel by physico-chemical reactions. Attachment becomes irreversible through specific molecular interactions between adhesins and ligands. Pioneer species influence the subsequent pattern of colonization by changing the local environment and by providing receptors for intra- and intergeneric coaggregation. The diversity of the microflora increases with time through waves of bacterial succession, so that eventually a climax community with a high species diversity is reached. This diversity is due, in part, to the development of food chains among the resident organisms. Climax communities are characterized by a degree of stability (microbial homeostasis); this stability is sustained by antagonistic and synergistic microbial interactions. As a consequence of differences in environmental conditions, the climax community varies at individual sites on a tooth surface.

On occasions, microbial homeostasis breaks down, and disease can occur. The most common disorders are caries and periodontal diseases which result from a shift in the balance of the resident oral microflora. Dental caries is the localized dissolution of the enamel or the root surface by acid produced from the microbial fermentation of dietary carbohydrates. Numerous epidemiological and laboratory studies have strongly implicated mutans streptococci in lesion formation; lactobacilli may be involved with lesion progression. If the dentine or pulp becomes infected, then a range of organisms can be isolated. Attempts to prevent caries involve the mechanical removal of plaque, and the use of fluoride, antimicrobial agents, fissure sealants, non-fermentable sweeteners, and vaccines. Periodontal diseases are a range of conditions in which the supporting tissues of the teeth are attacked. Tissue destruction follows plaque accumulation and involves the production of enzymes and cytotoxins, particularly by obligately anaerobic Gram-negative species, and the induction of a host inflammatory response. Treatment involves plaque removal and control, which on occasions may be augmented by antimicrobial agents.

Many oral micro-organisms are potential opportunistic pathogens. If the host defences are suppressed or if these organisms gain entry to sites not normally accessible to them, they can cause abscesses, actinomycosis, or candidosis. Dental procedures regularly produce transient bacteraemias which may result in infective endocarditis in patients with predisposing conditions.

Cross infection is also a potential hazard with many dental procedures, especially as frank pathogens (particularly viruses) can be carried in serum or saliva.

The mouth is a highly dynamic ecosystem. The microflora is held in balance through a series of interactions between the host and the resident micro-organisms. This balance contributes to the health of the individual by preventing colonization by exogenous species (colonization resistance). Care has to be taken to prevent this balance from being irreversibly perturbed.

Glossary

Several scientific disciplines are involved in the study of oral microbiology. Inevitably, specialist terminology will have been used in this book that is unfamiliar to students of different subjects. This glossary has been provided to help overcome these problems. Specialist terms are given a simple explanation in relation to their usage in this book. Such explanations should not be regarded as strict definitions.

Abscess a collection of pus

Acidogenic acid-producing (usually used to describe an organism producing sufficient acid to be capable of playing an active role in the demineralization of enamel and cementum, i.e. dental caries)

Aciduric acid-tolerating (acid-loving)

Acquired pellicle film of polymers of mainly salivary origin that are absorbed onto the tooth surface immediately following tooth cleaning

Actinomycosis infection due to *Actinomyces* spp

Adhesin molecule on the surface of a micro-organism involved in adhesion

Aetiology association of specific factors (including micro-organisms) with the cause of a disease

Aggregation adherence of bacterial cells to each other

Allogenic succession bacterial succession influenced by factors of non-microbial origin, e.g. tooth eruption

Approximal surface between adjacent teeth (Figure 2.2)

Atrophy shrinkage in size of an organ or tissue by reduction in size of its cells

Autochthonous population a characteristic member of the microbial community of a habitat

Autogenic succession bacterial succession influenced by microbial factors, e.g. the metabolism of pioneer species lowers the redox potential during plaque development. This allows obligate anaerobes to colonize

Bacteraemia micro-organisms present in the blood stream

Bacterial succession pattern of development of a microbial community (Figure 4.1)

Candidosis infection with *Candida* spp

Cariogenic dental caries-inducing (e.g. bacterium, diet, etc.)

Climax community stable complex microbial community that develops by, and is the final product of, the process of bacterial succession (Figure 4.1)

Coaggregation the attachment of a cell to a pre-attached organism by specific molecular interactions

Colonization resistance the ability of the resident microflora to prevent colonization by exogenous species (Table 4.7)

Commensalism an inter-bacterial interaction beneficial to one population but with a neutral effect on the other

Competition rivalry among bacteria for growth-limiting nutrients

Cryptitope a receptor on a host molecule for a microbial adhesin that is exposed only under certain conditions, e.g. when adsorbed to a surface or after enzyme cleavage

Cytotoxic therapy therapy that kills dividing cells, usually used to treat cancers

Demineralization dissolution of enamel or cementum by acid

Dental caries localized dissolution of the enamel or root surface by acid derived from the microbial degradation of dietary carbohydrates

Dental plaque tenacious deposit on the tooth surface comprising bacteria, their extracellular products and polymers of salivary origin

Fissures narrow grooves found mainly over the biting (occlusal) surface of a tooth (Figure 2.2)

Gingival crevice protected habitat formed where the teeth rise out of the gum (Figures 2.1 and 2.2)

Gingival crevicular fluid serum-like exudate bathing and flushing the gingival crevice. It has a considerable influence on the ecology of this region by introducing (1) nutrients for the microbial community and (2) components of the immune system

Gnotobiotic animal germ-free animal deliberately infected with a known bacterial population or microflora

Hepatitis inflammation of the liver

Hyperplasia increase in size of an organ or tissue by increase in the number of constituent cells

Hypertrophy increase in size of an organ or tissue by increase in size of its cells

Immunocompromised the state of being susceptible to infection by virtue of impairment or malfunction of the immune system

Infective endocarditis infection of the lining of the heart (endocardium)

Ligand host receptor for microbial adhesins

Metastasis spread through blood stream or lymphatic system to another organ or tissue

Microbial homeostasis the natural stability of the resident microflora of a site

Microbial taxonomy study of the classification of micro-organisms according to their resemblances and differences

Minimum infective dose the minimum number of micro-organisms required to cause an infection

Mitral valve valve between the left ventricle and atrium

Mucocutaneous affects both skin and mucous membranes

Necrosis death of tissues or cells

Needlestick injury puncture of the skin due to a sharp instrument

Niche the function or role of an organism in a habitat. Species with identical niches will, therefore, be in competition

Occlusal surface on the top of the tooth (Figure 2.2)

Opportunistic pathogen an organism normally enjoying a commensal relationship with the host but which has the potential to cause disease under extraordinary circumstances such as when the resistance of the host is reduced or when the organism is found in a new habitat

Osseointegrated the intregration of bone with artificial material

Osteomyelitis inflammation of bone caused usually by infection

Periodontal disease general term for several diseases in which the supporting tissues of the teeth are attacked

Periodontal pocket formed by the migration of the junctional epithelium at the base of the gingival crevice down the root of the tooth (Figures 2.1 and 7.1). The migration and subsequent tissue destruction is caused by a host inflammatory response to the microbial challenge and by the production of virulence factors (Table 7.9) by periodontopathogens

Periodontopathogen an organism implicated in the aetiology of periodontal diseases

Predisposing factor a condition or circumstance that makes an individual susceptible to a disease

Proto-co-operation an interbacterial interaction beneficial to all populations involved

Protonmotive force, pmf, a vectorial gradient of cations, principally protons, across a membrane that generates potential energy. This energy can be harnessed to ATP synthesis, motility, and transport of solutes

Pus a collection of micro-organisms, fluid and cellular debris

Redox potential, Eh, the oxidation-reduction potential of a site. Anaerobic bacteria prefer an environment with a low Eh

Resident microflora the microbial community associated with a particular habitat; this microflora usually lives in harmony with the host and has several beneficial functions to the host, e.g. colonization resistance

Retrovirus a virus that transcribes its RNA into DNA and back again. This is accomplished by the presence of a reverse transcriptase

Sialadenitis infection of the salivary glands

Siallogogue substance that encourages saliva production

Sialoliths stones in the salivary gland

Sinus a tissue tract or space lined with epithelium from which pus or fluids drain

Sjögren's syndrome a syndrome that involves dry mouths and eyes

Sub-gingival below the gingival (gum) margin, e.g. as in a sample taken from the gingival crevice or periodontal pocket (Figures 2.1 and 2.2)

Supra-gingival above the gingival (gum) margin (Figure 2.2)

Thrombus a blood clot

Vegetations blood clots on the heart lining or endocardium

Xerostomia dryness of the mouth usually due to impairment of salivary gland function

Index